TESTING REGRESSION MODELS BASED ON SAMPLE SURVEY DATA

My children Maria and Osamah Bhatti

Testing Regression Models Based on Sample Survey Data

M. ISHAQ BHATTI
School of International Business Relations
Faculty of Asian and International Studies
Griffith University

with a foreword by Maxwell L. King

Avebury

Aldershot · Brookfield USA · Hong Kong · Singapore · Sydney

Published by
Avebury
Ashgate Publishing Limited
Gower House
Croft Road
Aldershot
Hants GU11 3HR
England

Ashgate Publishing Company
Old Post Road
Brookfield
Vermont 05036
USA

British Library Cataloguing in Publication Data

Bhatti, M. Ishaq
  Testing Regression Models Based on
  Sample Survey Data
  I. Title
  519.536

  ISBN 1 85628 642 8

Library of Congress Catalog Card Number: 94-73470

Typeset by Jennilyn Mann
Griffith University
Nathan
Queensland 4111
Australia

Reprinted 1997

Printed in Great Britain by
Biddles Short Run Books, King's Lynn

Contents

List of tables

ix

xi

Foreword

Maxwell L. King

In a World that is increasingly becoming competitive, deregulated, instantaneous, computer-rich and data-rich, best practice demands greater use of statistics in decision-making in commerce and government. Two important tools are the sample survey and the linear regression model. Cost considerations often mean that survey data is gathered in blocks and sometimes subblocks of observations that may be influenced or affected by common factors. The best examples concern geographical groupings. Here subjects share a common environment including climate, prices, information, infrastructure, culture and so on. When such data is used to estimate a regression model, these latter influences, which individually may not be important, can add up to create unwelcome correlations in the regression errors.

In this book, M Ishaq Bhatti systematically explores the problem of testing for these unwelcome correlations in the linear regression model. He attacks the problem of test construction from the point of view that only the most powerful test will do. Whether lots of extra computing is required is not an issue. The computer revolution has continuously moved the bounds on what is a feasible amount of computing. It is also clear that these bounds will continue to expand in the future. The forward in Statistics, therefore, is to ask, as Dr Bhatti has done, what procedures would we want to use if computing power were not an issue.

Maxwell L. King
Monash University

Preface

This book considers several aspects of hypothesis testing problems associated with the linear regression model applied to multi-stage sample-survey data. Sample-surveys are increasingly being used as a tool by economists, businessmen and statisticians who need to gather data before making an important decision or giving much needed advice. There is an increasing tendency to perform regression analysis using sample-survey data in Market Research, Econometrics, Social and Biological sciences, Geology and Geography. The statistical universe in these fields, as indeed in many others, while conceived as a whole is often comprehended as constituted of a number of natural blocks. In such cases, it is more useful as well as procedurally more practical to collect data in terms of blocks. For example, the population of a country falls naturally into the states, cities, districts, subdistricts, or urban and rural blocks. Similarly a single industry is often perceived as a complex of blocks and subblocks, each are concerned with well defined processes of manufacture.

When regression analysis is used to analyse data obtained from such blocks, subblocks or groups, the residuals are often found to be correlated within ultimate blocks. This correlation is sometimes constant across blocks and its coefficient is called the intrablock or equicorrelation coefficient. This equicorrelation in the error term is also called a block effect. The main aim of this book is to develop and investigate tests for these effects.

This book is divided into three parts. Part one reviews the literature and recomends the use of a model-based approach in the analysis of survey data. It suggests a unified regression model which could be equally useful for survey statisticians and sometimes for panel-data econometricians. It also summarises the theory of point optimal testing and suggests how these optimal tests can be applied to the unified regression model in the subsequent chapters. In chapters three and four the small-sample properties of optimal tests for equicorrelation in standard symmetric multivariate normal (SSMN) and symmetric multivariate normal (SMN) distributions are developed and investigated respectively. The SMN distribution is a multivariate distribution in which all components of a k dimensional random vector, y, have equal means, equal variances and all covariances between components take the same value. These common covariances give rise to a common equicorrelation coefficient, ρ. In this part of the book, some optimal tests for two- and three-stage SSMN and SMN distributions for detecting intrablock and intrasubblock equicorrelation among blocked/subblocked observed random vectors are developed and investigated.

Part two extends part one to linear regression models in which the disturbances follow two-, three- and higher-stage SMN distributions. In this part of the book, an attempt is made to find an 'optimal' exact test for the intrablock equicorrelation for the two-stage linear regression model and then generalized to a multi-stage linear regression model by using King's (1987b) point optimal invariant (POI) approach. The possibility of the locally most mean powerful invariant (LMMPI) tests by using King and Wu's (1990) approach is also explored.

In part three of the book Cox and Solomon's (1986, 1988) model has been considered and the possibility of constructing a locally best invariant (LBI) test, POI test and LMMPI test is considered. Throughout this book the principle of invariance has been used to eliminate the nuisance parameters where possible, thus reducing the dimension of the testing problem.

Acknowledgements

Whatever merits of this book may be, they are in large measure due to my teachers at B.Z. University, Multan, Pakistan, The University of Alberta, Edmonton, Canada and lastly Monash University, Australia. Particularly, Professor Maxwell King, I owe a great deal for his help, invaluable guidance, continuous encouragement and useful discussions during the preparation of this book.

I also appreciate the patience and understanding of my wife who has experienced many lonesome moments while I was laboring over many details, torn between getting the text into print and enjoying her company.

Finally, I would like to express my appreciation to Jennilyn Mann not only for typesetting the manuscript but also contacting me via E-mail around the globe for the preparation of this book in a camera ready format. I am grateful to P. Segal and J.B. Ofosu for reading the manuscript.

Any errors that exist are solely mine and I trust there will be few.

1 Introduction

Preamble

Sample-surveys are now routinely being used as a tool by economists, business consultants, and statisticians who gather data and interpret it before making an important decision or giving much-needed advice. With the greater use of the computer, the statistical analysis of such data is becoming more sophisticated. Often, among others, linear regression methods are used to analyse the data. This is frequently done without due regard being given to the complications that can arise. For example these data sets may be subject to heterogeneity. It is common for such data to be collected in blocks or clusters for economic reasons. In this case, data elements from different blocks or clusters may be affected by any number of diverse effects such as climate, soil, prices, etc. The similarities within a block or cluster can give rise to what is known as 'intrablock or intracluster correlation' or simply 'the block effects' in the error terms of regression models. If these effects are strong enough, they make standard regression techniques inappropriate.

The main aim of this book is to develop and investigate diagnostic tests for these effects. Just as a doctor can run diagnostic tests on a patient to see if there is a specific medical problem, the econometrician/statistician can run a test on the regression model to see whether intrablock or intracluster correlation is present. The simplest form of intrablock correlation arises from two-stage cluster sampling. There are three-stage and multi-stage cluster sampling procedures as well. In a national survey the largest blocks might be, for example, states, the next largest blocks might be regions, then perhaps postal districts and finally streets. This is an example of four-stage cluster sampling. Because of state law and practices, one might also expect intrablock correlation within states. Diagnostic tests for intrablock correlation within smaller blocks such as e.g., regions and/or postal districts have been developed in this book and are evaluated using computer simulation techniques.

Generally, it has been common for researchers to ignore the existence of such block effects, arguing perhaps that the amount of these effects are very small and thus are unlikely to affect the analysis seriously. As pointed out by Holt, Smith and Winter (1980), and followed by Holt and Scott (1981), Deaton and Irish (1983), Moulton (1990) and more recently Bhatti (1994) this small amount of residual equicorrelation can cause the standard errors of the ordinary least squares (OLS) estimator to be biased downward seriously. The bias of the standard errors can result in spurious findings of statistical inference and hence leads to grave misleading conclusions. This book investigates the small sample power properties of some diagnostic tests for detecting the existence of such block effects from the multistage linear regression model.

The theory of statistical hypothesis testing enables us to develop such diagnostic tests. No matter whether these tests are classical, namely the Wald, likelihood ratio (LR) and Lagrange Multiplier (LM) or they are, Cox's non-nested tests (Cox, 1961 and 1962), Hausman's specification tests (Hausman, 1978), White's information matrix tests (White, 1982), conditional moment tests (Newey, 1985 and Tauchen, 1985) and/or point optimal tests (King, 1987b), our ultimate concern is with the test's power. This is mainly because the power of a test determines its accuracy and the chance of making a correct decision by incorporating any knowledge of the signs of the parameters under test.

Among the classical tests, the LM test is the most popular mainly because of its ease of calculation, but its size and power can be distorted seriously in small samples. Therefore, our alternative choice is to construct a test which ensures certain known optimal power properties. Examples of tests already available include locally best (LB) tests (Neyman and Pearson, 1936 and King and Hillier, 1985), the point optimal (PO) tests (King, 1987b) and Locally Most Mean Powerful (LMMP) tests (King and Wu, 1990). These tests are reviewed in detail in the next chapter. But in the light of their names (and claims) one would do well to note that each of these tests has really a small-sample optimal power property. Ideally, wherever the testing strategy is undertaken and small sample optimal tests are conducted, it is well worth comparing the power properties obtained through them with those obtained by large-sample tests (e.g. LM). At a practical level, the actual size and power properties yielded by small sample optimal tests should suggest whether a large sample test is warranted.

Deaton and Irish (1983) considered a linear regression model based on a two-stage clustered sampling design and suggested the use of the one-sided

LM (LM1) test whereas King and Evans (1986) showed that the test is LB invariant (LBI) for testing block effects. Recently, SenGupta (1987) considered a model based on standard symmetric multivariate normal (SSMN) distribution which, as will be seen in this book, is a special case of King and Evans model. He proposed a LB test for testing positive values of the equicorrelation coefficient for this model. Based on these studies the following issues seem worthy of further consideration.

1   How should PO tests based on SSMN and symmetric multivariate normal (SMN) distribution models for the same testing problem considered by SenGupta be developed? How do LMMP tests be constructed when the number of parameters being tested is more than one?

2   Is it possible to construct PO invariant (POI) tests for the same testing problem considered by King and Evans, if so, how well do they perform relative to other existing tests? Can LMMP invariant (LMMPI) tests be constructed for testing multiparameter extensions of the King and Evans' model?

3   How well does the LMMPI test perform for testing multi-block effects while dealing with the three-stage model? How well do the POI and the LM tests perform when the number of parameters being tested is more than one, in relation to the three-stage model? How can POI and LMMPI tests be constructed for the multi-stage model?

4   Is it possible to construct optimal tests for testing serial correlation coefficients for data comprised of large numbers of short samples?

This book attempts to answer these questions in the subsequent chapters. It should be noted that the principle of invariance has been used throughout this book to eliminate the nuisance parameters, wherever necessary, in order to reduce the dimensions of the various testing problems under consideration.

An outline of the book

In chapter two, the historical background of sample surveys and their associated problems, and a comparison between design-based versus model-based approaches is presented. A unified regression model is developed which shows that the modelling of block effects in survey data is

3

similar to that of individual effects in panel data. It is noticed that with a minor difference in concept and a change of notation one can easily use both models as substitutes for each other. In search of uniformally most powerful (UMP) tests, this chapter also reviews some of the existing tests and discusses the new optimal tests which make use of computer technology to maximize and/or improve their power under the alternative parameter space.

In chapter three the problem of testing for non-zero values of the equicorrelation coefficient of SSMN distributions is considered. The LB, Beta Optimal (BO) and LMMP tests for the case of two-stage SSMN distribution model is constructed.

Next the power of the two versions of the BO test with that of SenGupta's (1987) LB and the power envelope (PE) is compared and it thereby demonstrate the superiority of the BO tests. This chapter extends the two-stage SSMN model to a three-stage model and, when more than one parameter are being tested, constructs PO and LMMP tests. It also discusses the more complicated matters such as testing subblock effects in the presence of main block effects in the case of the three-stage SSMN model.

In chapter four analogous tests (i.e. POI and LMMPI tests) for the two- and three-stage SMN distributions are developed. The findings in this chapter suggest that the POI test for the case of the two-stage model is UMP invariant (UMPI). The most attractive features of both the POI and the LMMPI tests are that their power and critical values are obtainable from the standard F distribution. For the three-stage SMN model a POI test for testing subblock effects in the presence of main block effects is developed. The subsequent two chapters investigate various aspects of statistical inference in the linear regression model with disturbances that follow SMN distributions.

In chapter five, the size and small sample power properties of the numerous statistical tests associated with the two-stage linear regression model for detecting simple intrablock equicorrelation are investigated. The results suggest that the POI test is marginally better than the LM1 test for small and moderate sample sizes. Both the LM1 and the POI tests are approximately UMPI for some selected block sizes. It is also noted that the LMMPI test for testing the hypothesis for zero equicorrelation against the alternative that the correlation may differ from block to block is equivalent to that of the LM1 test. An empirical power comparison of the LM1, two-sided LM (LM2), Durbin-Watson (d), modified d (d*), POI tests and the PE shows the relative strengths of these tests.

Chapter six extends the model of chapter five to cover a more general situation and constructs POI and LMMPI tests for zero equicorrelation over different blocks (ρ_1) and over different subblocks (ρ_2). It also discusses how close the power of the LMMPI test is to that of the PE. The problem of

4

testing subblock effects, in the presence of main block effects is also explored in this chapter and the small sample power properties of the different statistical tests associated with the three-stage linear regression model are discussed. Monte Carlo studies are used to investigate the power performances in small samples. This chapter also generalizes these tests for the case of multi-stage linear regression models and develops some optimal tests.

Chapter seven generalizes Cox and Solomon's (1988) model and it suggests optimal tests for testing serial correlation coefficient in large numbers of small samples. The final chapter contains some concluding remarks.

2 Modeling survey data

Introduction

In recent years, there has been an increasing use of regression analysis based on survey data in empirical econometrics, statistics, market research, social, physical and biological-sciences.

Often the data arising from these areas are naturally in blocks. When regression is applied to data obtained from such blocks, subblocks or multiblocks (i.e. from multi-stage sample design), the regression errors can be expected to be correlated within ultimate blocks or subblocks. As one would expect, ignoring such correlation can result in inefficient estimators, and in seriously misleading confidence intervals and hypothesis tests. Part of the growing literature on the subject is based on the traditional sampling frame work such as, for example, the works of Konijn (1962) and Kish and Frankel (1974). Others such as Scott and Holt (1982), Deaton and Irish (1983), King and Evans (1986), Christensen (1986, 1987a, 1987b), Moulton (1986, 1990) and more recently, Dorfman (1993) and Rao et. al. (1993) have focussed on the model based approach.

This book follows a model-based approach in order to deal with the regression analysis problems of sample survey data. The literature survey of the model-based approach work reveals that most of the above mentioned authors have considered only the simplest form of intrablock correlation arising from the two-stage cluster sampling. In subsequent chapters, an extension of their work in two directions is explored. Initially, a PO test for testing intrablock correlation in a two-stage model so as to achieve comparability with the other existing tests is developed. Next having found these testing procedures superior in terms of power performance, it is used on a three-stage model to check further upon its viability and practicality.

In real life situations, sometimes, the two-stage models are woefully inadequate. For example, in an investigation of a national population with a basic unit of household, a two-stage model would be well-neigh useless. What is needed is a model that allows for blocking and subblocking at the levels of regions, states, wards, postal districts, streets or similar. In other words, a multi-stage model is needed. Hence, this book considers the use of a three-stage model in order to demonstrate further viability of our testing procedures and to understand the consequences of such an extension. Furthermore, the use of this model to a general multi-stage model is explored.

The main purposes of this chapter are to develop a unified linear regression model, review the theory of optimal hypothesis testing and to survey several existing diagnostic tests for detecting block effects associated with our model. It starts with an account of the history of the sample-surveys and their use in regression analysis. The choice of using the model-based approach when working with survey data is justified. Problems associated with blocked data are also raised. A unified regression model is developed which can be used by statisticians dealing with survey data and suggest that its special case can also be used by econometricians dealing with time series cross sectional and/or panel data. This section of the chapter therefore discusses the variety of models which involve fixed and random coefficients models with one and two-way error components models and finally the multi-stage linear regression model. The theory of optimal testing and discussion on some problems associated with the theory of PO testing are summarized. The latter part of the chapter reviews some articles which construct exact tests for block effects from SSMN, SMN and two-stage linear regression models. In the final section some concluding remarks are made.

Historical background of sample surveys

The idea of sample-surveying was developed by Kiaer (1895) and was adopted by the International Statistical Institute in its first meeting in Berlin in 1903. Next, Bowley (1906) supplied the theory of inference for survey samples. Later, Neyman (1934) was concerned with the choice between alternative sampling procedures for survey data, such as stratified, purposive and cluster sampling. Others who contributed significantly were Cochran (1942) who dealt with ratio and regression estimators; Hansen and Hurwitz (1949) who developed the selection procedures with probability proportion to size; Madow and Madow (1944) and Cochran (1946) considered the method of systematic selection.

Due to its wide application in different fields of human endeavour, quite a number of textbooks have been devoted to the subject of sample-surveys. Among the notable texts, Yates (1949) was the first, followed by Deming (1950), Cochran (1953), Hansen, Hurwitz and Madow (1953), Sukhatme (1954) and Kish (1965). Their latest editions are largely refinements and improvements on their original work although they contain new areas of thought which has generated further research in the area. For example, Godambe (1966), Royall (1968, 1970), Hartley and Rao (1968, 1969), Scott and Smith (1969) and Särndall (1978) have used the idea of finite and superpopulation in relation to the case of multistage and other survey designs.

Some even more interesting work has been done in the last few decades on the method of data analysis of complex sample surveys. Particularly in the field of regression analysis the study by Kish and Frankel (1974) and Pfefferman (1985) represent a new approach. They investigated the impact of an intrablock or equicorrelation coefficient on the regression analysis by using the standard sample-survey frame work. They pointed out that, if the intra-block correlation is ignored, then the estimates of the calculated standard errors of the regression coefficients underestimate the true standard errors.

Through the 1980's the practitioners and theoreticians have discussed the controversial issue of whether a design-based or a model-based approach should be adopted for regression analysis of sample survey data. The difference between the model-based linear least squares prediction approach and the design-based sampling distribution approach is discussed in the next section.

Model-based versus design-based approach

It is well known among survey statisticians that the design-based approach in regression mean as a relationship between the dependent variable, y and the independent variable, x, where x is obtained from well defined clusters or blocks of a finite population. In this approach, each observation is weighted by the reciprocal of its probability of being included in the sample in order to make an inference about the finite population on the basis of the observed sample. Here unbiased estimators of the intercept and the slope coefficient are obtained by taking the expectation over all possible samples and a consistent estimator is obtained by using Cochran's (1953) definition.[1] In the design-based approach, it is believed that the allocation of weights in the sample selection procedure made the inference about population

9

parameters inefficient and hence it does not lead to best linear unbiased estimators (BLUE).

In contrast to the design-based approach, the model-based approach always assumes that there exists a data generating process for the variables and therefore there is not a finite population. For example, when money demand, GDP or any other macroeconomic variable are observed, there is no finite population. This leads to using a stochastic model to explain the data generating process, as correctly as possible. The simplest stochastic model is

$$y_i = \alpha + \beta x_i + \varepsilon_i$$

where the ε_i's are i.i.d. $(0, \sigma^2)$. If the x_i's are exactly known, the OLS estimator is the minimum variance linear unbiased estimator (or BLUE) for β and α.

In this model, the word 'unbiased' refers to the expectations of overall possible realizations of the stochastic process while 'consistent' is used in the usual econometric or statistical sense.[2]

An interesting evaluation of the model-based and design-based approaches is given by Royall and Cumberland (1981). The authors compare the variance estimates of both approaches. Their empirical results favour the use of the model-based approach due to its efficiency. A further advantage of the model-based approach is that it can be applied using existing software regression packages. Examples of such computer packages are SURREGR (Holt, 1977), SUPER CARP (Hidiroglou, Fuller and Hickman, 1980) and PC CARP (Fuller, Kennedy, Schnell, Sullivan and Park, 1986).

This book, has adopted the procedure of using the model-based approach for the analysis of survey data.

Problems of blocked data

Scott and Holt (1982) considered the effects of two-stage design on the OLS estimator, particularly on efficiency and on standard errors. They showed that the OLS estimator is consistent but inefficient and the standard estimator for the variance-covariance matrix is consistent only if $\rho = 0$. Kloek (1981), Greenwold (1983) and Moulton (1986) have analyzed the magnitude of the bias. They have shown that the magnitude of the downward bias for the standard errors increases with the increase in the average block size, the intra-block or equicorrelation of the disturbances, and the equicorrelation of the regressors.

Despite these problems, survey data would always be preferred by the researchers due to their availability from secondary sources and its time and cost advantages over the time series cross-sectional and/or panel data sets. Because in hightech and competitive markets, the analysis of survey data is very important, the need for diagnostics tests on the validity of such models is a basic requirement for researchers. Discussion on some existing tests and proposed optimal tests in this area are detailed later in the chapter. Before this, next section presents a discussion on a unified linear regression model which would be used throughout this book in a different format.

A unified[3] regression model for survey data

Here a unified regression model mean an equation in which the grouping/blocking of homogeneous characteristics is reflected in the disturbance term, u_{ij}.

Two-stage linear regression model

Here in this model it is assumed that n observations are available from a two-stage sample with m blocks or clusters. Let m(i) be the number of observations from the ith block so that $n = \sum_{i=1}^{m} m(i)$. Then the simplest linear regression model of this form would be

$$y_{ij} = \sum_{k=1}^{p} \beta_k x_{ijk} + u_{ij} (i = 1,2,...,m, \quad j = 1,2,...,m(i)), \tag{2.1}$$

in which i is the block identifier, j is the observation identifier in the given block, β_k are unknown coefficients and x_{ijk} for k=1,2,...,p are observations on p independent variables, the first of which is a constant. It is assumed that u_{ij}'s are independent between blocks but equicorrelated within blocks. Hence

$E(u_{ij}) = 0$ and

$$\text{cov}(u_{ij}u_{st}) = \begin{cases} \sigma^2 & (i = s, j = t), \\ \rho\sigma^2 & (i = s, j \neq t), \\ 0 & (i \neq s). \end{cases} \tag{2.2}$$

Here ρ in (2.2), is the equicorrelation coefficient of the disturbances.

11

Models of this form have been considered by Fuller and Battese (1973), Fuller (1975), Campbell (1977), Holt and Scott (1981), Scott and Holt (1982), Deaton and Irish (1983), Christensen (1984), King and Evans (1986), Honda (1989) and more recently Bhatti (1994), among others. Detailed discussion on this model (and its simplest form) will be given in chapter five.

If it is assumed that $x_{ijk} = 0$ or $\beta_k = 0$, then the model (2.1) will become Bhatti's (1995) SMN distribution model. The further assumption of $\sigma^2 = 1$ in (2.1) through (2.2) make this as a SSMN distribution model (see Bhatti and King, 1990). The definitions and examples of these distributions are given in the next section.

Standard symmetric multivariate normal distribution

The purpose of this subsection is to define SSMN and SMN distributions and then investigate some of their applications in the field of statistics and econometrics. It is important to note that these distributions arise naturally in biometrics, education, genetics, psychology and related areas. Examples include the analysis of missing observations in time series (Sampson, 1976, 1978), generalized canonical variable analysis (SenGupta, 1983) and distributional modelling of repeated failure time measures (Crowder, 1985).

Recently, they have been applied in econometrics and statistics literature, for example, by SenGupta (1987, 1988), Williams and Yip (1989), Bhatti and King (1990) and Bhatti (1995). To econometricians and statisticians, the best known model based on these distributions is a special case of pooled time series and cross-section data. Another important application is in the regression analysis of data drawn from two (or higher) stage cluster or blocked surveys. For example, when census data are used, a block may be defined by a standard city block or an irregularly shaped area with identifiable political or geographical boundaries or it can be a group of industries or occupations etc. For the latter type of data, observations are correlated within ultimate clusters. This correlation is called the intrablock, intraclass, equi-, uniform or familial correlation coefficient, ρ. Ignoring such correlation leads to seriously misleading confidence intervals and hypothesis tests based on inefficient OLS estimates, (e.g., see Walsh (1947) and Halperin (1951)) and it also produces inefficient forecasts. In the next section these distributions are defined.

The SMN distribution is a multivariate normal distribution in which all the components of a k-dimensional random vector, y, have equal means, equal variances and all covariances between components take the same value (see Rao, 1973, p. 196). These common covariances give rise to a common correlation coefficient, ρ, which is called an equicorrelation coefficient. Thus, if $E(y) = \mu$ then SMN model can be written as

$$y \sim N(\mu, \sigma^2 \textstyle\sum(\rho))$$

where[4]

$$\textstyle\sum(\rho) = \begin{pmatrix} 1 & \rho & \rho & \dots & \rho \\ \rho & 1 & \rho & & . \\ \rho & \rho & 1 & & . \\ . & & & . & \\ . & & & & . \\ \rho & \rho & \rho & \dots & 1 \end{pmatrix}. \tag{2.3}$$

Similarly, a k-dimensional random vector y can be said to follow a SSMN distribution if it follows a SMN distribution and additionally its components' means and variances are zero and unity, respectively, i.e., $E(y) = 0$ and $\sigma^2 = 1$ and hence SSMN distribution model can be expressed as

$$y \sim N\left(0, \textstyle\sum(\rho)\right)$$

where $\sum(\rho)$ is given by (2.3). Though the literature on SMN distributions is quite extensive, no test for ρ has been proposed for SSMN distributions except SenGupta's (1987) LB test and Bhatti and King's (1990) beta optional test. Discussion on this test will be given later in this chapter.

Sampson (1976, 1978) has considered theoretical applications of SSMN distributions and developed the simple best asymptotic normal (BAN) estimation procedure for autoregressive, moving average and intraclass or equicorrelated models. He notes that SSMN distributions arise naturally from multivariate models in which means and variances of individual variables are known, thus allowing these variables to be standardized. Such standardizations are always made and play important roles in the techniques for reduction of dimensionality, e.g., in canonical variables (Anderson, 1984) and generalized canonical variables analysis (SenGupta, 1981, 1983).

This distribution is of interest to both theoreticians and practitioners for several reasons. One practical application of the SSMN distribution occurs when there are many observations on the individual variables but, because of historical, financial or practical reasons, there are comparatively few sets of joint observations. The individual observations can be used to obtain excellent estimates of means and variances which allow one to proceed as if these estimates are the true values. Such practical examples can be found in time series analysis, analysis of missing observations, psychometrics, generalized canonical variables and in biometrics. For some related results on models which follow this distribution, One may refer to Wilkes (1946), Sirivastva (1965), or more recently, SenGupta (1987, 1988). Another example of the use of such a distribution can be found in Crowder (1985) who gives a distributional model for repeated failure time measurements. Further, the SSMN distribution provides a practical example of a curved exponential family[5] and illustrates some associated difficulties and techniques concerned with inference on them, particularly when dealing with the testing of hypotheses.

One-way error component model

The model (2.1), under (2.2) is similar to that of the random effects model or one-way error component model used by econometricians in the analysis of panel data.
A simple (re)formulation appropriate in this case is

$$y_{it} = \sum_{k=1}^{p} \beta_k x_{itk} + u_{it}, \ (i = 1,2,...,N, \ t = 1,2,...,T)$$

where

$$u_{it} = \mu_i + v_{it}, \tag{2.4}$$

in which i=1,2,...,N, where N stands for the number of individuals (e.g. households) in the sample and t=1,2,...,T, where T stands for the length of the observed time series. Each of the μ_i's, (i=1,...,N) are called an individual effect and v_{it} is the usual (white noise) error term. In this reformulation (at this stage) it is assumed that every block has the same number of observations (T).

The basic outline of the model (2.4) has been drafted by the pioneers Balestra and Nerlove (1966), Wallace and Hussain (1969) and Maddala (1971). They assume that

14

1   The random variables μ_i and υ_{it} are mutually independent.

2    $E(u_{it}) = 0$. This implies that $E(\mu_i) = 0$ and $E(\upsilon_{it}) = 0$

3    $var(\mu_i) = \begin{cases} \sigma_\mu^2, & \text{for } i = i' \\ 0, & \text{otherwise} \end{cases}$

4    $var(\upsilon_{it}) = \begin{cases} \sigma_\upsilon^2, & \text{for } i = i', \ t = t' \\ 0, & \text{otherwise.} \end{cases}$

In comparing (2.1) with (2.4), it is noted that $u_{it} = \mu_i + \upsilon_{it}$, $\sigma^2 = \sigma_\mu^2 + \sigma_\upsilon^2$, and $\rho = \sigma_\mu^2 / \sigma^2$.

The only difference between (2.1) and (2.4) is that in model (2.4) the ith block consists of the time-series of the ith individual and the number of observations in a 'block' is T, the length of the time series. In the econometrics literature, this model is also called the one-way error component model. Useful references include Hsiao (1986), Honda (1989), Moulton and Randolph (1989), Baltagi and Li (1991) and Körösi, Mátyás and Székely (1992). This model is frequently being used to model panel data in the econometrics literature. In chapter five this model is discussed in detail and it has been called as the two-stage linear regression (2SLR) model.

Three-stage linear regression model

This subsection extends the 2SLR model of the previous subsection to more general situations where each block is divided into subblocks and then one can use the error component to capture these block/subblock effects to form a three-stage cluster sampling design model.   The three-stage linear regression (3SLR) model is expressed as,

$$y_{ijk} = \sum_{\ell=1}^{p} \beta_\ell x_{ijk\ell} + u_{ijk} \quad (i = 1,...,m, \ \ j = 1,...,m(i), k = 1,...,m(i,j))$$

(2.5)

in which i is the block identifier, j is the subblock identifier in the ith block, k is the observation identifier in the jth subblock of the ith block such that

$$n = \sum_{i=1}^{m} \sum_{j=1}^{m(i)} m(i,j),$$

where β_ℓ are unknown coefficients and $x_{ijk\ell}$ for $\ell=1,...,p$ are observations on p independent variables, the first of which is a constant. It is assumed that u_{ijk} are normally distributed with mean zero and its variance-covariance structure is given by

$$\mathrm{cov}(u_{ijk}\, u_{rst}) = \begin{cases} 0, \text{ (for } i\neq r, \text{ and } j,s,k \text{ and } t) \\ \rho_1\sigma^2, \text{ (for } i=r, \ j\neq s, \text{ any } k,t) \\ (\rho_1+\rho_2)\sigma^2, \text{ (for } i=r, \ j=s \ k\neq t) \\ \sigma^2, \text{ (for } i=r, \ j=s, \ k=t) \end{cases} \qquad (2.6)$$

where, $k, t = 1, 2, ..., m(i,j)$, $j, s = 1, 2, ..., m(i)$ and $i, r = 1, 2, ..., m$. Here in (2.6), ρ_1 and ρ_2 measure the main block and subblock effects, respectively. Note that if it is assumed that $\rho_2 = 0$, then the 3SLR model (2.5) will become equivalent to 2SLR model of the form (2.1). Further note that if it is assumed that each subblock in a main block has only one observation then at least mathematically one can easily show that (2.6) will become equivalent to that of the two-way error component model.

The general layout of our 3SLR model will be discussed in chapter six, where some diagnostic tests for detecting these block and/or subblock effects were constructed. Monte Carlo methods are used to compare the power performance of these tests with that of the LM tests.

Random coefficients model

Note that a model of the form (2.1) can also be expressed in terms of a random coefficients model where it is assumed that the regression coefficient, β_k varies from block to block. A formulation appropriate to the random regression coefficient is given by Bhatti (1993) and Bhatti and Barry (1994). The formulation of Bhatti (1993) is different from Maddala's (1977, p. 326) variance components model and Swamy's (1970)[6] and Hsiao's (1986)[7] random coefficients models.

In developing uniform regression models of the form (2.1) and (2.5) the most popular approach of error components models has been used, to model survey data. The general popularity of this approach among survey statisticians and panel data econometricians (see, Mátyás and Sevestre, 1992) can be attributed to the following facts:

1   They treat huge data bases as well as modest ones with equal ease.
2   Estimation and hypothesis testing methods arc derived from the classical well-known procedures.

3   The problems and difficulties presented remain within the traditional framework and hence they are well understood.

4   The theoretical frontiers are much more explored comparatively than for the other possible approaches.

5   The estimation and hypothesis testing results can be interpreted easily.

6   Most commonly used econometrics and statistical software packages can be applied to these models with only minor modifications.

The use of error component models in panel data was first suggested by Balestra and Nerlove (1966) and its use in survey data was pointed out by Holt, Smith and Winter (1980). Others to have proposed its use, include Wallace and Hussain (1969), Amemiya (1971), Swamy and Arora (1972), Fuller (1975), Fuller and Battese (1973, 1974), Campbell (1977), Rao and Kleffe (1980), Baltagi (1981), Scott and Holt (1982) and Dickens (1990). Just recently, Baltagi and Raj (1992), and Mátyás and Sevestre (1992) reviewed the various estimation and testing procedures in the context of panel data. As is noted above, modelling of panel data is very similar to that of the survey data. Therefore, one can apply nearly the same estimation procedures to the 2SLR or (in special situations) 3SLR models that those used in panel data modelling in order to obtain efficient estimates of the unknown parameters.

The hypothesis testing problems associated with the error components models of the type SSMN, SMN, 2SLR, 3SLR and the multistage linear regression models are the main concern of this book. In the next section the theory of point optimal testing and the problems and the difficulties involved in developing these tests will be summarized.

Optimal hypothesis testing

In hypothesis testing, one would always prefer to use a UMP test which maximizes the power curve over the entire parameter space. Unfortunately,

17

the existence of a UMP test does not happen very often in practice, perhaps only in special circumstances. When no UMP test exists, it is difficult to decide which is the preferred test as no single test can dominate in terms of power over the whole parameter space.

Apart from this, statisticians, econometricians and other social scientists are often faced with a high degree of frustration in their respective fields of hypothesis testing because of the limited amount of data available. Furthermore, the requisite data cannot always be generated or controlled by laboratory experiment. Hence with a small amount of information on hand, when testing for an economic theory (or a hypothesis in any other field), it is important to choose some optimal testing procedure which will provide a powerful test.

Cox and Hinkley (1974, p. 102) consider various techniques for constructing tests of simple null hypothesis, H_0, against a composite alternative hypothesis, H_a, when no UMP test exists. One of the techniques they consider is using the most powerful test against a specific alternative $\theta_a \in H_a$ as the test of H_0 against H_a. King (1987b) has labelled this technique the PO approach. This approach is reviewed later in this chapter. In future chapters the method of constructing optimal tests for the SSMN, SMN, 2SLR, 3SLR and the multi-stage models are considered. In the literature, this method has been shown to work well in a wide variety of testing problems provided that the specific alternative is chosen carefully. For example, a sequence of papers published by King (1982, 1983, 1985a, 1985b and 1987b) examines the performance of PO tests in testing for first order moving average regression disturbances (MA(1)), first-order autoregressive disturbances (AR(1)) and AR(1) against MA(1) disturbances in the linear regression model with small samples. Further, Evans and King (1985) and King and Skeels (1984) investigate the small sample power properties of PO tests for heteroscedastic disturbances and joint AR(1) and heteroscedastic regression disturbances, respectively.

These studies viewed together, make us draw the conclusion that certain point optimal tests have more desirable small sample power properties than others of its competitors, including the LMMP tests. These tests will be defined in somewhat more detail later.

Over the past twenty years, the advanced technology provided by the computer revolution has brought enormous benefits to statisticians, econometricians, biometricians and other researchers working in the quantitative methods areas. With the ever cheapening cost of computing operations and the greater capacity of computer memory, optimal testing is becoming more viable, popular and its value is recognized in research work. No matter whether the hypothesis is simple or composite, one-sided or two-sided, with or without nuisance parameters, advances in computing technology have helped us to solve these problems. However, this book, is limited to only one-sided hypothesis testing problems because the block/subblock effects under tests are mostly positive.

Review of the theory of point optimal testing

To begin by considering a general form of hypothesis testing, i.e. testing

H_0: y has density $f(y, \omega)$, $\omega \in \Omega$

against

H_a: y has density $g(y, \phi)$, $\phi \in \Phi$

in which y is an observed $n \times 1$ vector, ω is a $j \times 1$ vector and ϕ is an $i \times 1$ vector. It is assumed that the possible range of parameter values have been determined by all given knowledge in order to keep the parameter spaces, Ω and Φ, as small as possible.

Generally speaking there are three different cases in hypothesis testing which are defined as follows:

Case (1): This is a problem of testing simple null against a simple alternative hypothesis, with ω_1 and ϕ_1 fixed and known as parameter values such that $\Omega = \{\omega_1\}$ and $\Phi = \{\phi_1\}$, i.e.,

H_0^1: y has density $f(y, \omega_1)$,

against

H_a^1: y has density $g(y, \phi_1)$.

Therefore, the Neyman-Pearson Lemma (see, Lehmann, 1959, p.65) implies that rejecting H_0^1 for large values of

$r = g(y, \phi_1) / f(y, \omega_1)$

is a MP test.

Case (2): This is a problem of testing simple null hypothesis, H_0^1 against the composite alternative, H_a. If r is used as a test statistic, then this test, by

19

construction, is MP in the neighbourhood of $\phi = \phi_1$. The critical value for this test can be found by solving

$$\Pr\{r > r' | y \text{ has density } f(y, \omega_1)\} = \alpha$$

for r' where α is the desired significance level.

Case (3): It is probably the most practical case and involves testing a composite null hypothesis, H_0, against the composite alternative, H_a. In this situation, it may be difficult to construct an MP test in the neighbourhood of $\phi = \phi_1$. This is because the size of the test and the power of the test may depend on some unknown values of the parameters, i.e.,

$$
\begin{aligned}
\text{Size of Test} \quad &= \Pr\{y \in W | H_0 \text{ is true}\} \\
&= \int_W \text{pdf}_0(y, \omega) dy \\
&= \alpha(\omega), \text{ say.}
\end{aligned}
\tag{2.7}
$$

Note that $\alpha(w)$, in (2.7) may depend on the unspecified part of ω where W is the critical region. Similarly,

$$
\begin{aligned}
\text{Power of Test} \quad &= \Pr\{y \in W | H_a \text{ is true}\} \\
&= \int_W \text{pdf}_a(y, \phi) dy \\
&= \beta(\phi), \text{ say.}
\end{aligned}
\tag{2.8}
$$

Here also note that $\beta(\phi)$ in (2.8), may depend on ϕ and hence the power of the test may not always be large for some values of ϕ. In equations (2.7) and (2.8), pdf_0 and pdf_a are the probability density functions under the null and alternative hypothesis, respectively.

The standard approach suggested by Lehmann and Stein (1948) is to control the maximum probability of a type I error by the choice of critical value. This is done by solving

$$\sup_{\omega \in \Omega} \Pr\{r > r^* | y \text{ has density } f(y, \omega)\} = \alpha \tag{2.9}$$

for r^* which is the critical value. In general, the choice of r' is greater than r^*. Note that if Ω is a closed set, then

$$\Pr\{r > r *|y \text{ has density } f(y,\omega)\} = \alpha \qquad (2.10)$$

will be true for at least one $\omega \in \Omega$. If there is a value for ω_1 which leads to $r'=r^*$, then the critical regions of the two tests correspond. Thus, it can be said that the test based on r is an MP test in the neighbourhood of $\phi = \phi_1$ of H_0 against H_a.

If in the case where the ω_1 value exists and it is also known then constructing a PO test will be much simpler. Therefore, it is important to check for the existence of ω_1 before conducting a PO test, i.e., one has to check whether or not there exists ω_1 value such that (2.10) holds for $\omega = \omega_1$ subject to the constraint that (2.9) also holds. King (1987b, p.175) has outlined the method in calculating the left hand side of (2.10) with the help of readily available computer subroutine packages.

The iterative procedure he mentioned can be summarised as follows:

1   Choose a fixed value of ω_1.

2   Solve (2.10) to find r^* and then check to see if (2.9) holds. If it does hold, then the values of ω_1 and r^* have been found. If it does not hold, another choice of the ω_1 value is needed by moving it towards the ω values which caused the violation in (2.9). Then repeat step two.

Unfortunately, in practice it is not always so simple. An ω_1 value sometimes might not exist in a particular testing problem. Lehmann (1959, p.90-94) developed an approach for overcoming the non-existence of an ω_1 value, that is, to reduce a composite null hypothesis to a simple one by considering the weighted average of the distribution over the parameter space. Thus, H_0 is replaced by H_λ with the density function of y is given by

$$h_\lambda(y) = \int_\Omega f_\omega(y)d\lambda(\omega).$$

Lehmann assumes that λ is the degenerate distribution at $\omega = \omega_1$. Under this assumption with the size α, if the most powerful test of a simple null hypothesis, H_λ, against a simple alternative, H_a, has a value of its size less than or equal to α with respect to H_0, then the resultant test is also a PO test of H_0 against H_a. However, if this assumption does not hold, Lehmann's approach is not appropriate. King (1987b, pp. 179-180) and Silvapulle

21

(1991) consider an approximate point optimal (APO) testing approach to solve this problem.

For instance, in case (3) above, a test statistic r with its critical value r* determined by (2.9) is said to be an APO test if $\Pr\{r > r * | H_0^1\}$ is close to α, say within 5% of α. This is because the distribution of H_0 has been changed such that this probability was equal to α then our test would be a PO test. Therefore, there is a need to look for an ω_1 value which minimizes the difference between α and $\Pr\{r > R * | y$ has density $f(y, \omega_1)\}$ h,

i.e. $\alpha - \Pr\{r > r * | y$ has density $f(y, \omega_1)\}$. $\qquad(2.11)$

In addition, since (2.11) also depends on ϕ_1, it is worthwhile to consider varying ϕ_1 to minimize the measure of approximation in (2.11) further. Recently, King (1989) and Silvapulle and King (1991) have used the APO approach for testing AR(4) regression disturbances in the presence of AR(1) and testing MA(1) against AR(1) disturbances, respectively. Brooks (1993) has used an APO test for testing for a single Hildreth-Houck random coefficient against the alternative that the coefficient follows the return to normalcy model.

Invariance approach and optimal tests

Apart from these, one of the most common problems involved in hypothesis testing is the existence of nuisance parameters which may be present under both the null and alternative hypotheses. The construction of a PO or an APO test, requires searching for an appropriate ω_1 value in the parameter space Ω. If the parameter space Ω can be restricted to be as small as possible, then it will be helpful in finding such value for ω_1. Obviously, one should check whether theoretical considerations can indicate some possible restrictions which can be imposed on the range of various parameters under both H_0 and H_a.

In general, there are two different approaches which help to solve this problem. They are the 'similar approach' and the 'invariance approach'. A rejection region of size α is called a 'similar region' of size α if for all values of the nuisance parameters the probability of rejecting the null hypothesis is exactly equal to α when the null hypothesis is true. Hillier (1987) observed that there is an exact equivalance between similar tests and invariant tests of H_0. The 'invariance approach' has had considerable success in the application of PO tests because invariance arguments can be used to eliminate nuisance parameters. The reasoning behind this approach is that if an hypothesis testing problem is invariant to a class of transformations on the observed sample, it is desirable that the test procedure also have this property. For example, as is seen in the subsequent chapters in relation to 2SLR, 3SLR and multi-stage linear regression models that changing of the scale of y and adding a known linear combination of regressors to the rescaled y still preserves the family distributions under both the null and the alternative hypotheses. The conclusion is that the invariance transformation may change the numerical values of the parameters but the functional form of the likelihood function still remains the same. Just recently, Ara and King (1993) have shown that the tests based on maximal invariant (i.e. invariant tests) and tests based on the marginal likelihood procedures are equivalance.

As is known in hypothesis testing, for the purpose of test construction a sample space can be divided into two regions, namely, a rejection region and an acceptance region. An invariant test is one for which each pair of points in the sample space that can be related by a transformation either both fall into the rejection region or they fall outside the rejection region. Therefore, decisions on the problem of how to partition the sample space can be simplified into choosing which sets of points related by transformations shall be in the rejection region and which sets shall be out of it.

A convenient summary statistic called a maximal invariant tends to solve the problem of deciding which set of points are to be in a certain region. A maximal invariant exhibits the property of invariance and every other invariant statistic can be written as a function of it. Hence, only the problem based on the maximal invariant rather than on the observed sample needs to be dealt with. Hopefully, this way the distributions under the null and the alternative hypotheses will have fewer parameters than in the original testing problem. However, the main disadvantage of the invariance arguments is that it is often difficult to find the distributions of the maximal invariant under null and under the alternative hypotheses. Nevertheless, throughout this book, invariance principle would be used in deriving the PO and the locally optimal tests.

Optimality criterion

A remaining question is, how does one choose a point at which the power is optimized? In the situation when no UMP test exists, Cox and Hinkley (1974, p.102) suggest three different approaches in the choice of points:

1   to pick 'somewhat arbitrary "typical" point' in the alternative parameter space and use it in the test in order to find a point optimal solution;

23

2   to remove this arbitrariness choose a 'point' which is very close to the null hypothesis in order to maximize the power locally near the null hypothesis; and

3   to choose a test which maximizes some weighted average of the powers.

Option (1) is known as the 'point optimal' solution discussed earlier in this chapter, and reviewed in some detail by King (1987b). In that paper, King notes that a class of PO tests is also called 'beta optimal' if its power function is always a monotonic nondecreasing function of the parameter under test and it reaches a predetermined value of the power, say, p_1, most quickly. The concept of beta-optimality was first introduced by Davies (1969), where he optimized power at a value of 0.8, which seems to give good overall power. The construction of the BO test in chapter three optimizes power at a predetermined level of power $p = 0.5$ and 0.8.

In contrast to PO and/or BO tests, option (2) leads us to a locally most powerful test often also known as a LB test, a term preferred throughout this book. The LB test is also optimal in the sense that its power curve has the steepest slope at H_0 of all power curves from tests with the same size.

The LB solution has been proposed by Neyman and Pearson (1936) and followed by Fergusson (1969), Efron (1975), King and Hillier (1985) and SenGupta (1987) among others. Let it be assumed that there is interest in testing $H_0: \theta = 0$ based on y which has been defined earlier as an $n \times 1$ random vector whose distribution has probability density function $f(y|\theta)$ where θ is a $p \times 1$ vector of unknown parameters. When $p=1$, the LB test of $H_0: \theta = 0$ against $H_a^+: \theta > 0$ is that with critical regions of the form

$$\left. \frac{\partial \ell n f(y|\theta)}{\partial \theta} \right|_{\theta=0} > c_1 \tag{2.12}$$

where c_1 in (2.12) is a suitably chosen constant; see Ferguson (1967, p.235) and King and Hillier (1985). King and Hillier (1985) noted that this test is equivalent to the LM1 test based on the square-root of the standard LM test statistic. Against the two-sided alternative $H_a: \theta \neq 0$ when $p=1$, Neyman and Pearson (1936) proposed a test which has the critical regions of the form

$$\left. \frac{\partial^2 f(y|\theta)}{\partial \theta^2} \right|_{\theta=0} > c_2 f(y|\theta) \left. \right|_{\theta=0} + c_3 \left. \frac{\partial f(y|\theta)}{\partial \theta} \right|_{\theta=0} \tag{2.13}$$

and which yield LB unbiased (LBU) tests where the constants c_2 and c_3 are chosen so that the critical region has the required size and is locally unbiased. The critical regions of the form (2.13) were labelled as type A regions by Neyman and Pearson in their (1936) paper. They also proposed type A_1 tests which are known as uniformly most powerful unbiased (UMPU) tests. Neyman (1935) showed how to construct type B and type B_1 tests, namely, LBU and UMPU tests, respectively.

For the higher dimension parameter spaces, i.e., where $p \geq 2$, the LBU critical region of $H_0 : \theta = 0$ against $H_a : \theta \neq 0$ is obtained by using Neyman and Pearson (1938) type C LBU regions. These regions have constant power on each of a given family of concentric ellipses in the neighbourhood of the null $\theta = 0$ and this constant power is maximized locally. Isaacson (1951) introduced type D regions to rectify this objection (also see Lehmann, 1959, p.342). As Wu (1991) pointed out that the type D regions are obtained by maximizing the Gaussian curvature of the power function at $\theta = 0$. Further he added, 'in practice, type D critical regions need to be first guessed and then verified'. SenGupta and Vermeire (1986) introduced the class of LMMP unbiased (LMMPU) tests which maximize the mean curvature of the power hypersurface at the null point, i.e. $\theta = 0$ within the class of unbiased tests. Their critical region is of the form

$$\sum_{i=1}^{p} \frac{\partial^2 f(y|\theta)}{\partial \theta_i^2}\bigg|_{\theta=0} > c_0 f(y|\theta)\bigg|_{\theta=0} + \sum_{i=1}^{p} c_i \frac{\partial f(y|\theta)}{\partial \theta_i}\bigg|_{\theta=0}$$

where the constants c_j, $j=1,2,...,p$, are chosen such that the test has the nominated size and is locally unbiased.

King and Wu (1990) suggested a multivariate analogue of the one-sided testing problem, i.e. $H_0 : \theta = 0$ against $H_a^+ : \theta_1 \geq 0,...,\theta_p \geq 0$ with at least one strict inequality. They showed that LMMP test of H_0 against H_a^+ has the critical region

$$s = \sum_{i=1}^{p} \frac{\partial \ell n f(y|\theta)}{\partial \theta_i} > c. \tag{2.14}$$

They also noted that there is always the possibility that one may be able to find such a test of the form (2.14) which is LB in all directions from H_0 in the p-dimensional parameter space. Neyman and Scott (1967) call this property 'robustness of optimality' while King and Evans (1988) call such tests uniformly LB (ULB). Like UMP tests, ULB tests may not always exist. In such cases, one may wish to consider a weaker optimality criterion. The aim of the next section is to explore the options of constructing PO, BO, LB and LMMP critical regions in relation to unified regression model for testing block/subblock effects.

Existing tests for block effects

This section begins by considering articles that construct exact tests for block effects i.e., whether the equicorrelation coefficient between observations within a block is zero, $H_0:\rho=0$, against an alternative that it has a positive value, $H_a:\rho>0$. This section is divided into two parts. The first part reviews articles with the stacked random vectors of different observations which follow SSMN and SMN distributions. The second part will concentrate on testing for block and subblock effects in the context of the multi-stage linear regression model, when disturbances follow only SMN distributions.

SSMN and SMN models

Sampson (1976, 1978) has considered applications of the SSMN distribution and the estimation of its equicorrelation coefficient, ρ. William and Yip (1989) has used the idea of ancillary statistics on the two-stage SSMN distribution model. They found that the tests and the confidence intervals based on ancillary statistics are conditionally optimal. In this section, the main concern is to survey articles which have considered exact optimal tests so that the power of the new and the existing exact tests can be compared.

Recently, SenGupta (1987, 1988) proposed a LB test for non-zero values of the ρ in a two-stage SSMN distribution model (2.3). He showed that in the context of (2.3), the likelihood ratio test that ρ takes a given value has a number of theoretical and practical shortcomings. He constructed the LB test of H_0: $\rho = 0$ against H_a: $\rho > 0$ and found that his tests do not share these shortcomings. This test reject H_0 for small values of

$$ d = \sum_{i=1}^{m}\sum_{j=1}^{k} y_{ij}^2 - \sum_{i=1}^{m}\left(\sum_{j=1}^{k} y_{ij}\right)^2 . \tag{2.15} $$

Empirical power comparisons of these tests showed the superiority of the LB test. A LB test whose statistic is of the form d in (2.15) can be viewed as a test which optimizes power in the neighbourhood of the null hypothesis.

In chapters three and four PO tests are considered which optimize power at a predetermined point away from the null hypothesis. The main findings of these chapters are that for the testing problem in the context of SSMN and SMN distributions, the point optimal tests are BO and UMP invariant (UMPI), respectively. An extension of the two-stage SSMN and SMN models to multistage model has been considered and a class of new optimal tests has been proposed.

Multi-stage linear regression models

Recently, Honda (1991) prposed a standardised test for the two-way error component model and found that the size of his test improves the accuracy for the small sample sizes. Baltagi, Chang and Li (1991) have considered one-way and two-way error component models and proposed various tests on the variances-covariances of the individual and time effects. Mátyás and Sevestre (1992) have reviewed some of the tests suggested by Honda and Baltagi et. al which were associated with the panel data models. Others who have considered error component models are Honda (1985) and Moulton and Randolph (1989). They analysed the behaviour of the LM1 tests. They demonstrated that it is more powerful and more robust to non normal error terms than the LM2 test King and Hillier (1985) showed that the LM1 test is LBI. These properties suggest the usefulness of the LM1 test in the context of modelling survey data. The main objective of this section is to survey some of those articles which consider testing block/subblock effects associated with the 2SLR and the 3SLR models.

Deaton and Irish (1983) have considered the 2SLR model of the form (2.1). They discussed the problem of testing for block effects and suggested the use of the LM1 (score) test. This test rejects $H_0:\rho=0$ for large values of

$$g = \hat{\rho}\left(\frac{1}{2}\sum_{i=1}^{m}m(i)[m(i)-1]\right)^{\frac{1}{2}}$$

where $\qquad\qquad\qquad\qquad\qquad\qquad\qquad\qquad\qquad\qquad\qquad\qquad$ (2.16)

$$\hat{\rho} = \sum_{i=1}^{m}\sum_{j=1}^{m(i)}\sum_{k<j}\hat{u}_{ij}\hat{u}_{ik}\bigg/\left[\frac{1}{2}\sum_{i=1}^{m}m(i)\{(m(i)-1)\}\hat{\sigma}^2\right],$$

in which $\hat{u}_{ij}$ are OLS residuals and $\hat{\sigma}^2$ is the standard OLS estimator of the disturbance variance. It is worth noting that asymptotically g has a standard normal distribution under H_0. The two-sided Lagrange multiplier test rejects H_0 for large values of g^2, which has an asymptotic χ^2 distribution with one degree of freedom under the null hypothesis.

The g test in (2.16) requires the knowledge of the block structure for its application. Deaton and Irish (1983) have pointed out that this may not always be available. Block information is often removed from survey data before release, although the original block ordering is retained. Their suggestion, that in such cases one may have to resort to the use of the Durbin-Watson test, is attractive, especially since this test is available in most standard output regression packages.

King and Evans (1986) have also considered the 2SLR model with the same testing problem. They discussed the small sample optimal power properties of the LM1, LM2 tests and the DW test. They empirically compared the powers of these tests and found that the LM1 test is LBI. In chapter five when dealing with the 2SLR model this testing problem has been considered and the construction of the POI, LMMPI and King's (1981) modified DW test is worked out. It is found that the power of the POI test is marginally superior to that of the existing tests in the literature.

Others who considered the 2SLR model are, Wu et al. (1988) and Rao et al. (1993). They allow negative and/or unequal equicorrelation in their models. Wu et al. (1988) developed a standard F-test for testing H_0: $C\beta = b$ whereas Rao et al. (1993) have developed a generalised least square (GLS) F test. The power and size comparison between Wu et al's F test and Rao's et al's GLS F test demonstrates the superiority of the GLS F test. Just recently Wu and Bhatti (1994) have considered the 3SLR model and suggested the use of POI and LMMPI tests for testing block effects and misspecification in regression models based on survey data. This book will follow Wu and Bhatti's procedure of detecting block/subblock effects from the SSMN, SMN, 2SLR, 3SLR and finally the multistage Linear regression models in the subsequent chapters.

Concluding remarks

This chapter reviewed other scholars' work to understand the issues and problems connected with the use of sample surveys in regression analysis. Most of the scholars have considered the 2SLR model which was sometimes inappropriate in real life situations. Therefore, their work was extended to multi-stage model by incorporating block, subblock and multiblock effects in the error terms. The choice of using model-based approach in the analysis of survey data was explained and a unified regression model for survey and panel data was developed. While these are the preliminary concepts, a proper understanding of them is necessary before embarking on the task of developing the optimal tests. The theory of PO testing was summarized .

Articles which construct exact tests for testing block effects were reviewed so that the tests used in this book could stand alongside for comparison later on. The literature survey of the PO tests revealed that they performed pretty well in other situations in terms of their size and power properties as compared to asymptotic tests. This suggests that it is worthwhile to explore the power properties of the optimal tests, like PO, BO, LB and LMMP, in testing these block/subblock effects. Therefore, in the subsequent chapters tests on SSMM, SMN, 2SLR, 3SLR and multi-stage linear regression models are applied.

Notes

1. An estimator is called consistent in this approach if the estimate becomes exactly equal to the population value when n = N, that is, when the sample consists of the whole population.

2. A consistent estimator here is an estimator which converges in probability, as the sample size increases with a decreasing variance, to the parameters of which it is an estimator.

3. It unifies regession modelling of survey and panel data as it can be noted from equation 2.4.

4. The structure of the variance-covariance matrix of the observed vector, y, is identical to that of the residuals in panel and survey data, see Balestra and Nerlove (1966) and Bhatti (1994), respectively.

5. Efron (1975) considered arbitrary one-parametric families and attempted to quantify how nearly 'exponential' they are. Since it is well known in the literature that one parametric exponential families have very nice properties for estimation, testing and other inference problems, statistical curvature is identically zero for exponential families and positive for nonexponential families. Statistical curvature is closely related to Fisher (1925) and Rao's (1962, 1963) theory of second-order efficiency.

6. Swamy's (1970) model ignores the equicorrelation coefficient, ρ, within blocks and considers hetroscedasticity. Whereas in the model used in this study the variance terms of the diagonal elements of the ith block are constant within the block.

7. Hsiao (1986) ignores equicorrelation within blocks.

3 Some tests of the standard symmetric multivariate normal distribution

Introduction[1]

The literature survey of chapter two revealed that the model which follows the SSMN distribution has wide applications in the field of biometrics, econometrics, statistics, biological and physical sciences. The main aim of this chapter is to consider several aspects of hypothesis testing problems associated with the models based on two- and three-stage SSMN distributions. Particularly, discussion will be concentrated on developing the optimal tests discussed in the previous chapter. Among other things this chapter also explores the possibility of constructing a PO test for testing the subblock equicorrelation coefficient in the presence of main block equicorrelation.

The plan of this chapter is as follows. In the following section, the two-stage SSMN model is introduced and the BO test is constructed. Among other things, this section also discusses the LB test two versions of BO tests, the PE, and gives tables of selected critical values. An empirical power comparison involving LB and BO tests for the case of two-stage SSMN distributions is also reported. The application of King and Wu's (1990) LMMP test to more complicated models is given in the next section, where means and variances are known but equicorrelations are unknown and they vary from block to block is demonstrated. The main interest of that section is to test whether these varying equicorrelations are equal to zero against the alternative that they have some positive values, i.e., $H_0:\rho_i=0$, $i=1,...,m$ against $H_a:\rho_i\geq0$, $i=1,...,m$, with at least one inequality being a strict inequality. How LMMP and PO critical regions can be constructed in the context of the three-stage SSMN distribution model is examined. There is also discussion on the more complicated matters such as testing for subblock effects in the presence of main block effects in the case of the three-stage block model. The final section contains some concluding remarks.

Two-stage SSMN model and the tests

Let y be an observed $k \times 1$ random vector with a SSMN distribution, i.e.,

$$y \sim N\left(0, \sum(\rho)\right),$$

where the matrix $\sum(\rho)$ has already been defined by (2.3) in chapter two. Further, note that the simplest form of (2.3) can be written as

$$\sum(\rho) = \left[(1-\rho)I_k + \rho E_k\right], \tag{3.1}$$

where I_k is the $k \times k$ identity matrix and E_k is the $k \times k$ matrix with all elements equal to unity. Then the inverse of (3.1) is given by,

$$\sum(\rho)^{-1} = (1-\rho)^{-1}\left[I_k - \rho\{1+(k-1)\rho\}^{-1}E_k\right]. \tag{3.2}$$

Hence, the density function of y when $\sum(\rho)$ is nonsingular can be expressed as

$$f(y,\rho) = \frac{1}{(2\pi)^{k/2}\left|\sum(\rho)\right|^{1/2}} \exp\left[-\frac{1}{2}\left\{\frac{\left(\sum y_i^2\right)}{(1-\rho)} - \frac{\rho\left(\sum y_i\right)^2}{\{1+(k-1)\rho\}(1-\rho)}\right\}\right],$$

$$-\infty > y_i > \infty, i = 1,...,k, \text{ provided } \frac{-1}{(k-1)} < \rho < 1. \tag{3.3}$$

The restriction on the range of ρ is imposed to avoid the cases where $\sum(\rho)$ is singular. For large values of k, this range effectively becomes $0 \le \rho < 1$. Therefore the remainder of this chapter will be restricted to nonnegative values of ρ.

Let $y_1, y_2,...,y_m$ denote random samples of observed values from each of the m blocks, or states, or any other statistical entity from the distribution given in (3.3) and let Y denote the $n \times 1$ stacked vector of these independent random vectors where n=mk. It can be expressed in matrix form as follows:

$$Y = \begin{pmatrix} y_1 \\ y_2 \\ \vdots \\ y_m \end{pmatrix} = \begin{pmatrix} y_{11} \\ y_{12} \\ \vdots \\ y_{1k} \\ \vdots \\ \vdots \\ y_{m1} \\ y_{m2} \\ \vdots \\ y_{mk} \end{pmatrix}.$$

Now, the distribution of Y will be of the form

$$Y \sim N\big(0, \Delta(\rho)\big)$$

where

$$\Delta(\rho) = I_m \otimes \sum(\rho)$$
$$= (1-\rho)I_n + \rho D, \tag{3.4}$$

$\otimes$ denotes the Kronecker product and D is the nxn block diagonal matrix,

$$D = I_m \otimes E_k$$

$$= \begin{pmatrix} E_k & 0 & \cdots & 0 \\ 0 & E_k & \cdots & 0 \\ \cdot & & \cdot & \cdot \\ \cdot & & \cdot & \cdot \\ \cdot & & & \cdot \\ 0 & & \cdots & E_k \end{pmatrix}. \tag{3.5}$$

Furthermore, the inverse of $\Delta(\rho)$ is

$$\Delta^{-1}(\rho) = (1-\rho)^{-1}\Big[I_n - \rho\{1 + (k-1)\rho\}^{-1}D\Big]. \tag{3.6}$$

33

Note that the error covariance matrix in (3.4) involves only one unknown parameter, ρ, which ranges from 0 to +1. In the next subsection the BO test for detecting the presence of equicorrelation within blocks will be constructed.

Beta-Optimal test

The problem of interest is one of testing

$$H_0 : \rho = 0$$

against the one-sided alternative

$$H_a : \rho > 0.$$

Under H_0, the density function of Y is given by

$$f_0(Y) = (2\pi)^{-n/2} \ \exp\left\{\frac{-1}{2} Y'Y\right\}$$

while under H_a it is given by

$$f_a(Y;\rho) = (2\pi)^{-n/2} |\Delta(\rho)|^{-1/2} \exp\left[\frac{-1}{2} Y'\Delta^{-1}(\rho)Y\right].$$

Consider first the simpler problem of testing H_0 against the simple alternative hypothesis $H_1 : \rho = \rho_1 > 0$ where ρ_1 is fixed and known. The Neyman-Pearson Lemma implies that the critical region

$$|\Delta(\rho_1)|^{-1/2} \exp\left[\frac{-1}{2} Y'\Delta^{-1}(\rho_1)Y + \frac{1}{2} Y'Y\right] \geq c$$

or equivalently

$$r(\rho_1) = Y'\left(\Delta^{-1}(\rho_1) - I_n\right)Y < c_\alpha \tag{3.7}$$

is the most powerful test where c_α is an appropriate critical value.

It should be noted that under (3.6), one can express $r(\rho_1)$ in (3.7), as a simple series of sums as follows, i.e.

$$r(\rho_1) = (a* - 1)Y'Y - a*b*Y'DY$$

$$= (a* - 1)\sum_{i=1}^{m}\sum_{j=1}^{k}y_{ij}^2 - a*b*\left(\sum_{i=1}^{m}\sum_{j=1}^{k}y_{ij}\right)^2,$$

where

$$a* = \frac{1}{1-\rho_1}, \quad b* = \frac{\rho_1}{1+(k-1)\rho_1}.$$

In order to compute c_α, let $A = (\Delta^{-1}(\rho_1)-I_n)$, and let λ_i, $i=1,...,n$, be the eigenvalues of A and let P be an $n \times n$ orthogonal matrix such that

$$PAP' = \begin{pmatrix} \lambda_1 & 0 & . & . & . & 0 \\ 0 & \lambda_2 & . & . & . & 0 \\ . & & . & . & & . \\ . & & . & . & & . \\ . & & . & . & & . \\ 0 & & . & . & \lambda_n \end{pmatrix}.$$

Then under H_0

$$\Pr[r(\rho_1) < c_\alpha]$$

$$= \Pr\left[Y'AY < c_\alpha\right],$$

$$= \Pr\left[\sum_{i=1}^{n}\lambda_i\xi_i^2 < c_\alpha\right]$$

$$= \Pr\left[\frac{\rho_1}{1-\rho_1}\chi_{m(k-1)}^2 - \frac{\rho_1(k-1)}{\{1+(k-1)\rho_1\}}\chi_m^2 c_\alpha\right] \tag{3.8}$$

where $\xi = (\xi_1, \xi_2,...,\xi_n) \sim N(0, I_n)$ and χ_j^2 denotes a chi-squared random variable with j degrees of freedom. The last equality of (3.8) can be obtained by noting that the eigenvalues of D are k and zero with multiplicities of m

35

and m(k -1), respectively. This together with (ii) and (iii) of lemma 1, in appendix 3A implies that the eigenvalues of $\left(\Delta^{-1}(\rho_1) - I_n\right)$ are

$$\frac{-\rho_1(k-1)}{[1+(k-1)\rho_1]} \quad \text{and} \quad \frac{-\rho_1}{(1-\rho_1)}$$ with multiplicities m and m(k-1), respectively.

In order to find c_α such that (3.8) equals the required significance level, α, (3.8) can be evaluated using either Koerts and Abrahamse's (1969) FQUAD subroutine or Davies' (1980) algorithm. Alternatively, one can note that in (3.8), $r(\rho_1)$ is expressed as the weighted difference of two independent chi-squared random variables so that its probability density function is given by SenGupta (1987, Theorem 3).

For the wider problem of testing H_0 against H_a, the test based on $r(\rho_1)$ is most powerful at $\rho=\rho_1$ and is therefore a PO test. A central question is: how should ρ_1 be chosen? Strategies for choosing the point at which a PO test optimizes power are discussed in King (1987b). One approach is to choose a value of ρ_1 arbitrarily. Another is to take the limit of $r(\rho_1)$ tests as ρ_1 tends to zero. This results in SenGupta's LB test. This chapter will follow King's favoured approach of choosing ρ_1 so that the resultant test's power is optimised at a predetermined level of power denoted p_1. There seems little point in optimising power when it is very low as the LB test does or when it is one or nearly one. Optimizing power at middle power value say 0.5 or 0.8 is favored. In order to do this, the power of the test needs to be calculated readily.

Consider the Cholesky decomposition of $\Delta(\rho)$, namely

$$\Delta(\rho) = TT'$$

where T is a n x n nonsingular, lower triangular matrix. Then

$$\Delta^{-1}(\rho) = \left(T^{-1}\right)' T^{-1}$$

and under H_a

$$z = T^{-1}Y \sim N(0, I_n).$$

Because of the block diagonal nature of the $\Delta(\rho)$, T is also block diagonal of the form

36

$$T = \begin{pmatrix} S_k & 0 & . & . & . & 0 \\ 0 & S_k & . & . & . & 0 \\ . & & . & . & & . \\ . & & . & . & & . \\ . & & & . & & . \\ 0 & . & . & . & S_k \end{pmatrix}$$

where S_k is a kxk lower triangular matrix whose elements are obtained by

$$s_{ij}(\rho) = \rho \sqrt{\frac{(1-\rho)}{[1+(i-2)\rho][1+(i-1)\rho]}}$$

and

$$s_{ii}(\rho) = \sqrt{\frac{(1-\rho)[1+(i-1)\rho]}{[1+(i-2)\rho]}} \tag{3.9}$$

for

$i = 1,2,...,k; \ j = 2,3,...,k$ and $j>i$.

Thus for any critical value c_α, the power of the critical region $r(\rho_1) < c_\alpha$ is

$$\Pr\left[r(\rho_1) < c_\alpha | Y \sim N(0, \Delta(\rho)) \right]$$

$$= \Pr\left[z'T'(\Delta^{-1}(\rho_1) - I_n)Tz < c_\alpha \right]$$

$$= \Pr\left[\sum_{i=1}^{n} \lambda_i \xi_i^2 < c_\alpha \right], \tag{3.10}$$

where λ_i, $i = 1,2,...,n$, are the eigenvalues of

$$T'\left(\sum{}^{-1}(\rho_1) - I_n\right)T \text{ and } \xi_i \sim N(0,1).$$

37

It follows from lemma 2 of appendix 3A, that the eigenvalues of $T'\left(\Delta^{-1}(\rho_1) - I_n\right)T$ are also the eigenvalues of

$$\left(\Delta^{-1}(\rho_1) - I_n\right)\Delta(\rho)$$

$$= \frac{\rho_1(1-\rho)}{(1-\rho_1)}I_n + \frac{\rho_1\left[\rho\left\{(\rho_1-1)(k-1)+1\right\}-1\right]}{\left[(1-\rho_1)\left\{1+(k-1)\rho_1\right\}\right]}D$$

$$= aI_n + bD$$

say. The eigenvalues of the latter matrix are

$$a = \frac{\rho_1(1-\rho)}{(1-\rho_1)} \tag{3.11}$$

with multiplicity of mk-m and

$$a + bk = \frac{-\rho_1(k-1)\left\{1+(k-1)\rho\right\}}{\left\{1+(k-1)\rho_1\right\}} \tag{3.12}$$

with multiplicity of m. Thus $r(\rho_1)$ in (3.10) can also be expressed as a weighted difference of two independent chi-squared random variables, so (3.10) can be evaluated using any of the methods outlined above for calculating (3.8). In the special case in which power is being evaluated at $\rho = \rho_1$, (3.11) and (3.12) reduce to ρ_1 and $-(k-1)\rho_1$, respectively, so that (3.10) can be written as

$$\Pr\left[\rho_1\chi^2_{m(k-1)} - (k-1)\rho_1\chi^2_m < c_\alpha\right]$$

$$= \Pr\left[\chi^2_{m(k-1)} - (k-1)\chi^2_m < c_\alpha / \rho_1\right]. \tag{3.13}$$

Observe that both (3.11) and (3.12) decline in value as ρ increases. Given (3.10), this means that the test's power increases as ρ increases which implies that the test is a BO test. As noted by King (1987b, p.197-198), a PO test is BO if its power function is always a monotonic non-decreasing function of the parameter under test. The concept of beta-optimality was first introduced by Davies (1969) who suggested that the value of ρ_1 should be 0.8.

Given the desired level of significance, α, and the level of power, p_1, at which power is optimized, then ρ_1 and the associated critical value c_α, can be found by using (3.13) as follows:

1   Solve

$$\Pr\left[\chi^2_{m(k-1)} - (k-1)\chi^2_{m} < c_\alpha / \rho_1\right] = p_1$$

for c_α / ρ_1

2   Given the solution of the ratio of c_α / ρ_1 from (1) above, determine c_α and ρ_1 by solving

$$\Pr\left[r(\rho_1) < c_\alpha\right] = \alpha.$$

The remainder of this section will denote $r(\rho_1)$ test as the r_{ρ_1}.

Selected one and five percent significance points, c_α, and their associated ρ_1 values for the $r_{0.5}$ (i.e., $p_1=0.5$) and $r_{0.8}$ (i.e., $p_1=0.8$) tests are tabulated in tables 3.1 and 3.2 of appendix 3C, respectively. They were calculated using a FORTRAN version of Davies' (1980) algorithm. These calculations were made for vectors of size $k=2,3,....,10$ and for sample of sizes $m=5,10,15$ and 25.

Power comparison of BO and LB tests with PE

Point-optimal tests can be used to trace out the maximum attainable PE for a given testing problem. In this case, this can be done by evaluating the power of the $r(\rho_1)$ test at $\rho = \rho_1$ over a range of ρ_1 values. The PE provides an obvious benchmark against which test procedures can be evaluated. If the power of a test is always found to be close to the PE, then it can be argued that the test is an approximately UMP test. An example of such a finding is given by Shively (1988).

It is also of interest to compare the power curves of SenGupta's LB test with those of the two versions of the BO tests, $r_{0.5}$ and $r_{0.8}$, to see whether one of these tests is close enough to the PE to be called an approximately UMP test.

SenGupta has used Efron's (1975) criterion of statistical curvature, γ_θ, to measure the small-sample performance of the LB test. Based on 'very rough calculations', Efron (1975, p.1201) suggested that

$$\gamma_{\theta_0}^2 \geq 1/8$$

is a 'large' value where θ_0 is the value of the parameter under the null hypothesis. In this case it is reasonable to question the use of a LB test. For the testing problem defined in the previous section entitled 'Beta-optimal test' , SenGupta found that (3.14) is equivalent to mk $\leq$ 64. How good is Efron's rule in this case, dealing with SSMN distributions?

With these thoughts in mind, the PE has been computed and compared with the powers of the LB and the two versions of the BO tests, i.e., $r_{0.5}$ and $r_{0.8}$, at the five percent level of significance. Calculations of these powers at ρ=0.05, 0.1, 0.2, 0.3, .., 0.9 for m=10, 15, 25 and k=2, 3, 4, 6, 10 have been done. These calculations are obtained by using a FORTRAN version of Davies' (1980) algorithm. Selected results are tabulated in Tables 3.3, 3.4 and 3.5 of appendix 3C. The values for ρ=0.7 and 0.9 have been omitted because, especially for large values of m and k, they are very similar to those for ρ=0.8.

These results demonstrate that the PE and the powers of all tests increase as k, m and ρ increase, while other variables remain the same. As expected, of the three tests, the LB test is the most powerful for ρ values associated with small PE probabilities, whereas the $r_{0.5}$ test is most powerful for ρ values associated with middle PE probabilities and the $r_{0.8}$ test is most powerful for ρ values associated with large PE probabilities. Particularly for larger k and m values, the PE and all three power curves generally reach a value of one as ρ increases. The PE reaches this maximum value first, followed by the power curves of the $r_{0.8}, r_{0.5}$ and LB tests, respectively.

More importantly, for each combination of k and m values, the LB test always has the largest maximum power deviation below the PE of the three tests. For example, for k=2 and m=10,15,25, this maximum power difference is 0.246, 0.162 and 0.095, respectively. In contrast, the largest maximum power difference between the $r_{0.5}$ test and the PE is never greater than 0.031 while that for the $r_{0.8}$ test is never greater than 0.027. On the basis of these results it can be said that the $r_{0.5}$ and $r_{0.8}$ tests are approximately UMP, at least at the 5% level. If either k or m is increased, the maximum power difference from the PE decreases as either k (or m) increases while other variables remain the same.

40

Efron's rule of questioning the use of a LB test when (3.14), or equivalently mk $\leq$ 64 holds, appears to work well in this situation. Maximum deviations from the PE when mk $\leq$ 64 range from 0.246 to 0.071 while for mk > 64, they range from 0.057 to 0.015. There does seem to be a tendency for the rule to work better for smaller m values.

Finally, there is the question of which of the r_{p_1} tests is better. While the $r_{0.5}$ test has larger maximum deviations from the PE, the $r_{0.8}$ test has larger maximum percentage deviations. This is because the $r_{0.5}$ test has increased power for lower levels of power while the $r_{0.8}$ test is relatively more powerful for higher levels. The differences between the two tests are not great. The choice of test, therefore, boils down to a choice between extra power at lower or higher values of ρ.

A more complicated model and the LMMP test

The error covariance matrix in (3.4) involves only one parameter, ρ. But in practice it may be that the structure of the correlated observations changes from block to block or from group to group and hence the equicorrelation coefficient also varies from block to block or from industry to industry. Broadly speaking, one might view the error as being the sum of two components. The first is a disturbance for each block that is constant for each observation within the block. The second is a disturbance for individual observations. If the variance of this latter disturbance varies from block to block then one can obtain a correlation coefficient that also varies from block to block. For example, in studying the consumption patterns of different income groups in a particular locality, one can find that observations within each group may follow a similar pattern, but their consumption behaviour may vary from one group to another. Therefore, it is reasonable to assume that the degree of intragroup correlation also differs from one group to another group. Hence, this would result in different variances and covariances, which leads to different equicorrelation coefficients for different income groups.

In the literature these situations have been explored by many authors and the applications of various tests for the equality of means, variances-covariances and equicorrelation coefficients can be found in Wilks (1946), Votaw (1948), Geisser (1963), Selliah (1964) and Srivastava (1965). The main objective of the next section is to construct a one-sided test for testing whether these varying equicorrelation coefficients are equal to zero over different groups and even over different subgroups, against the alternative that they have some positive values by using Bhatti's (1992) LMMP and PO tests.

41

Locally most mean powerful test

The discussion in this section begins by noting that if different groups have different equicorrelation coefficients then the covariance matrix $\Delta(\rho)$, in (3.4) can be expressed as,

$$
\Delta(\rho) = \begin{bmatrix} \Sigma(\rho_1) & 0 & \cdots & 0 \\ 0 & \Sigma(\rho_2) & \cdots & 0 \\ \vdots & \vdots & \ddots & \vdots \\ 0 & \cdots & \cdots & \Sigma(\rho_m) \end{bmatrix}
$$

$$
= \bigoplus_{i=1}^{m} \Sigma(\rho_i),
$$

(3.15)

which is block-diagonal with submatrices of the form

$$
\Sigma(\rho_i) = (1 - \rho_i)I_k + \rho_i E_k
$$

(3.16)

for i=1,2,...,m, blocks, where $\rho = (\rho_1, \rho_2, ..., \rho_m)'$ and I_k and E_k have already been defined in the previous sections.

The problem of interest is one of testing

$$
H_0 : \rho_1 = \rho_2 = ... = \rho_m = 0
$$

against

$$
H_a : \rho_1 \geq 0, \ \rho_2 \geq 0, \ ..., \ \rho_m \geq 0,
$$

with at least one inequality being a strict inequality. This is an m-dimensional hypothesis testing problem. Therefore King and Wu's (1990) theorem to construct a LMMP test for testing H_0 against H_a can be used, in the case that the distribution of Y is,

$$
Y \sim N(0, \Delta(\rho)),
$$

where the $n \times n$ matrix $\Delta(\rho)$ is given by (3.15) with submatrices (3.16), and is such that $\Delta(0) = I_n$. The density function of Y with ρ being an $m \times 1$ vector is given by

$$f(Y|\rho) = (2\pi)^{-n/2}|\Delta(\rho)|^{-1/2} \exp\left[-\frac{1}{2}Y'\Delta^{-1}(\rho)Y\right]$$

$$\text{(3.17)}$$

where

$$\Delta^{-1}(\rho) = \begin{bmatrix} \Sigma^{-1}(\rho_1) & 0 & \cdots & & 0 \\ & & & & \cdot \\ 0 & \Sigma^{-1}(\rho_2) & & & \cdot \\ \cdot & & \cdot & & \\ \cdot & & & \cdot & \\ \cdot & & & & \\ \cdot & & & & \\ 0 \cdots & & \cdots 0 & & \Sigma^{-1}(\rho_m) \end{bmatrix},$$

$$\text{(3.18)}$$

such that the inverse of each of $\Sigma(\rho_i)$ is given by (3.2), i.e.,

$$\Sigma^{-1}(\rho_1) = (1-\rho_i)^{-1}\left[I_k - \rho_i\{1+(k-1)\rho_i\}^{-1}E_k\right],$$

From (2.13) of chapter two (i.e., King and Wu's, 1990, equation 3), the LMMP test of H_0 against H_a, based on the density (3.17), is given by the critical region

$$\sum_{i=1}^{m} \frac{\partial \ell nf(Y|\rho)}{\partial \rho_i}\bigg|_{\rho=0} > c,$$

$$\text{(3.19)}$$

where c is an appropriate constant chosen to give the required significance level of our test.

Thus, from (3.17), (3.18) and (3.19), the LMMP test is based on critical regions of the form

$$\sum_{i=1}^{m} -\frac{1}{2}\frac{\partial \ell n|\Delta(\rho)|}{\partial \rho_i}\bigg|_{\rho=0} - \sum_{i=1}^{m} \frac{1}{2}\frac{\partial(Y'\Delta^{-1}(\rho)Y)}{\partial \rho_i}\bigg|_{\rho=0} > c.$$

The first summation on the left is simply a constant. Evaluating the second summation gives the critical region

$$-\frac{1}{2}c_1 - \frac{1}{2}\sum_{i=1}^{m} Y'A_i Y > c,$$

or equivalently,

$$d = \sum_{i=1}^{m} d_i = \sum_{i=1}^{m} Y'A_i Y = Y'AY \leq c_2, \tag{3.20}$$

where $A = \sum_{i=1}^{m} A_i$ and c_2 is a suitably chosen constant. Note that the matrices A_i can be obtained from (3.18) as follows, i.e.,

$$A_i = \frac{\partial \Delta^{-1}(\rho)}{\partial \rho_i}\Big|_{\rho=0}, \quad i = 1, 2, ..., m$$

$$= \{a_{i,j\ell}\}$$

where

$$a_{i,j\ell} = \begin{cases} -1, & \text{for } (i-1)k < j < ik+1, \\ & (i-1)k < \ell < ik+1 \\ & \text{and } j \neq \ell; \\ 0, & \text{otherwise} \end{cases} \tag{3.21}$$

for $i=1,2,...,m$ and $j, \ell=1,2,...,n$. That is, the A'_is are block diagonal matrices whose blocks are $m \times m$ and all but the ith diagonal block which is $\left(I_k - E_k\right)$ are zero.

Thus by (3.16) and (3.18),

$$A = I_m \otimes \left(I_k - E_k\right)$$

$$= I_n - D,$$

where D is defined in (3.5). It is worth noting that (3.20) is a LB test of H_0 in the direction $\rho_1 = \rho_2 = ... = \rho_m > 0$, which is not unusual. A similar example of this type of result can be found while dealing with quarter-dependent fourth-order autocorrelated models (see King and Wu,

1990, p.17-18). Moreover, under (3.22), d in (3.20) can also be expressed in a simplified form as follows,

$$d = Y'AY$$

$$= Y'(I_n - D)Y$$

$$= \sum_{i=1}^{m}\sum_{j=1}^{k} y_{ij}^2 - \sum_{i=1}^{m}\left(\sum_{j=1}^{k} y_{ij}\right)^2,$$ (3.23)

which is in the form of 2.14 of chapter two (i.e., Sen Gupta's LB test). Hence Sen Gupta's LB test is also a LMMP test.

Further, one can obviously note that in (3.23), d is a quadratic form in normal variables, therefore its critical value c_2 may be obtained by evaluating probabilities of the form,

$$\Pr[d \le c_2] = \Pr\left[Y'(I_n - D)Y \le c_2\right]$$

$$= \Pr\left[\sum_{i=1}^{n} \lambda_i \xi_i^2 \le c_2\right]$$

$$= \Pr\left[\chi^2_{m(k-1)} - (k-1)\chi^2_m \le c_2\right]$$ (3.24)

where $\xi = (\xi_1,...,\xi_n)' \sim IN(0, I_n)$ and λ_i, $i=1,...,n$ are the eigenvalues of the matrix $(I_n - D)$, namely 1 and $(1-k)$ with multiplicities of $m(k-1)$ and m, respectively.

Three-stage SSMN model and the tests

Theory

This section extends the two-stage SSMN distributions model considered earlier, to the more general situation where observations are obtained from a three-stage (or higher-stage) block design. Suppose y_{ijk} denotes a kth (third-stage) selected variate value from a jth (second-stage) subblock of the ith (first-stage) main block. Let Y denote the nx1 random vector which is formed by stacking different observations for each subblock, then stacking different subblocks for each block and finally stacking for different blocks.

45

There are a total of n observations which are sampled from m first stage blocks (or groups), with m(i) subblocks from the ith block and with m(i,j) observations from the jth subblock of the ith block, such that

$$n = \sum_{i=1}^{m} \sum_{j=1}^{m(i)} m(i,j).$$

Then the random vector Y is said to follow the three-stage SSMN distribution model[2] if

$$Y \sim N\left(0, \Omega\left(\rho_1, \rho_2\right)\right) \tag{3.25}$$

where $\Omega(\rho_1, \rho_2) = \bigoplus_{i=1}^{m} \Omega_i(\rho_1, \rho_2)$ is block diagonal, with submatrices

$$\Omega_i(\rho_1, \rho_2) = \left(1 - \rho_1 - \rho_2\right)I_{m_i} + \rho_1 E_{m_i} + \rho_2 \bigoplus_{j=1}^{m(i)} E_{m(i,j)} \tag{3.26}$$

where $m_i = \sum_{j=1}^{m(i)} m(i,j), I_{m_i}$ is an $m_i \times m_i$ identity matrix, E_{m_i} and

$E_{m(i,j)}$ are $m_i \times m_i$ and $m(i,j) \times m(i,j)$ matrices, respectively, of ones.

Although unequal block and subblock sizes can easily be used in this model, but for the sake of convenience, in exposition, the rest of this chapter will consider only balanced case where data are balanced over subblocks. Therefore, it is assumed that $T = m(i,j), N_i = m(i)$, for $i=1,...,m$ and $j=1,...,N_i$, then (3.26) can be written as

$$\Omega_i(\rho_1, \rho_2) = \left(1 - \rho_1 - \rho_2\right)\left(I_{N_i} \otimes I_T\right) + \rho_1\left(e_{N_i} e_{N_i}' \otimes e_T e_T'\right)$$
$$+ \rho_2\left(I_{N_i} \otimes e_T e_T'\right) \tag{3.27}$$

where e_T and e_{N_i} are $T \times 1$ and $N_i \times 1$ vectors of ones.

Locally most mean powerful test

The error covariance matrix in (3.26) or equivalently in (3.27) involves two unknown parameters, i.e., the intrablock equicorrelation coefficient, ρ_1 and the intra-subblock equicorrelation coefficient, ρ_2.[3] As mentioned earlier in chapter one and two that ignoring such an intrablock or intra-subblock equicorrelation coefficient may lead to seriously misleading estimation and hypotheses testing results based on inefficient inference procedures.

Therefore, the problem of interest in this section is to test

$$H_0: \rho_1 = \rho_2 = 0 \tag{3.28}$$

against

$$H_a: \rho_1 \geq 0, \ \rho_2 \geq 0, \ \text{(excluding } H_0\text{)}. \tag{3.29}$$

For this problem the LMMP test is based on critical regions of the form (3.20), i.e.,

$$d^* = Y'AY \leq c_3, \tag{3.30}$$

where c_3 is a suitably chosen constant and in this case $A = A_1 + A_2$, in which

$$A_1 = \overset{m}{\underset{i=1}{\oplus}} A_{1i}$$

$$A_2 = \overset{m}{\underset{i=1}{\oplus}} A_{2i}$$

and by (3.27)

$$A_{1i} = I_{N_i} \otimes I_T - e_{N_i} e'_{N_i} \otimes e_T e'_T = \left(I_{N_i T} - E_{N_i} \otimes E_T\right),$$

$$A_{2i} = I_{N_i} \otimes I_T - I_{N_i} \otimes e_T e'_T = \left(I_{N_i T} - I_{N_i} \otimes E_T\right).$$

The rest of this section is constrained to the case of equal blocks, i.e., $N_i = s$, and also of equal subblocks, i.e., $m(i,j) = T$. Therefore d^* in (3.30), can be expressed as,

$$d^* = Y'A_1 Y + Y'A_2 Y$$

$$= Y'(I_n - D_1)Y + Y'(I_n - D_2)Y$$
$$= 2Y'Y - (Y'D_1 Y + Y'D_2 Y)$$

47

$$= 2\left[\sum_{i=1}^{m}\sum_{j=1}^{s}\sum_{k=1}^{T}y_{ijk}^2 - \left[\sum_{i=1}^{m}\left(\sum_{j=1}^{s}\sum_{k=1}^{T}y_{ijk}\right)^2 + \sum_{i=1}^{m}\sum_{j=1}^{s}\left(\sum_{k=1}^{T}y_{ijk}\right)^2\right]\right] \qquad (3.31)$$

where

$$A_1 = I_n - D_1,$$

$$A_2 = I_n - D_2,$$

$$D_1 = I_m \otimes E_{sT},$$

$$D_2 = I_{ms} \otimes E_T,$$

in which E_T and E_{sT} are TxT and sTxsT matrices of ones, respectively. Note that d* in (3.30) or in (3.31) is a quadratic form in normal variables, therefore, its critical value, c_3, may be obtained by evaluating the probabilities of the form,

$$\Pr[d* < c_3] = \Pr[2Y'Y - Y'(D_1 + D_2)Y < c_3]$$

$$= \Pr[Y'\{2I_n - (D_1 + D_2)\}Y < c_3]$$

$$= \Pr\left[\sum_{i=1}^{n}\lambda_i \xi_i^2 < c_3\right] \qquad (3.32)$$

where c_3 is a suitably chosen constant, $\xi = (\xi_1,...,\xi_n)' \sim N(0, I_n)$ and λ_i for i=1,...,n, are the eigenvalues of the matrix $\{2I_n - (D_1 + D_2)\}$, namely 2, (2-T) and $\{2-T(s+1)\}$ with multiplicities of ms(T-1), m(s-1) and m, respectively. Therefore (3.32) can be written as,

$$\Pr(d* < c_3) = \Pr\left[2\chi_{ms(T-1)}^2 + (2 - T)\chi_{m(s-1)}^2 + (2 - T(s+1))\chi_m^2 < c_3\right].$$

In the next section it will be explained how the PO test for testing (3.28) against (3.29) can be constructed, especially when dealing with the three-stage SSMN distribution.

Point optimal test

For the same testing problem discussed in the previous section, here King's (1987b) point optimal test is constructed which would be used to obtain the maximum attainable power, i.e. the PE over the alternative hypothesis parameter space. For any particular test its performance can be assessed by checking how close its power is to the PE. Here in this section the PO test statistic for testing $H_0: \rho_1 = \rho_2 = 0$ against the $H_a: \rho_1 > 0$, $\rho_2 > 0$ will be introduced.

Based on (3.25), choose $(\rho_1, \rho_2)' = (\rho_{11}, \rho_{22})'$ was chosen, a known point at which optimal power in the alternative parameter space is obtained. By following the procedure of the Beta-optimal test, with error covariance matrix (3.26), one can obtain the PO test for testing simple null hypothesis, $H_0: \rho_1 = \rho_2 = 0$ against a simple alternative hypothesis $H_a: \rho_1 = \rho_{11} > 0, \rho_2 = \rho_{22} > 0$, that is, to reject the null hypothesis for small values of

$$r\left(\rho_{11}, \rho_{22}\right) = Y'\left(\Omega^{-1}\left(\rho_{11}, \rho_{22}\right) - I_n\right)Y,$$

where

$$\Omega^{-1}\left(\rho_{11}, \rho_{22}\right) = \left[(1 - \rho_{11} - \rho_{22})I_n + \rho_{11}D_1 + \rho_{22}D_2\right]^{-1}.$$

This inverse can be evaluated by using Searle and Henderson's (1979) procedure. Therefore, $\Omega^{-1}(\rho_{11}, \rho_{21})$ is obtained as follows,

$$\Omega^{-1}\left(\rho_{11}, \rho_{22}\right) = (1 - \rho_{11} - \rho_{22})^{-1}\left[I_n - (1 - \rho_{11} - \rho_{22} + T\rho_{22})^{-1}\right.$$

$$\left.\left\{\rho_{11}(1 - \rho_{11} - \rho_{22})(1 - \rho_{11} + T\rho_{22} + sT\rho_{11})^{-1}D_1 + \rho_{22}D_2\right\}\right], \qquad (3.33)$$

where D_1 and D_2 are defined earlier by (3.31), i.e. $D_1 = \left(I_m \otimes E_{sT}\right)$ and $D_2 = \left(I_{ms} \otimes E_T\right)$.

Note that $r(\rho_{11}, \rho_{22})$ is a quadratic form in normal variables, therefore its critical value, c_α, may be obtained by evaluating the probabilities of the form,

49

$$\Pr\left[r(\rho_{11},\rho_{22}) < c_\alpha\right]$$

$$= \Pr\left[Y'\left(\Omega^{-1}(\rho_{11},\rho_{22}) - I_n\right)Y < c_\alpha\right]$$

$$= \Pr\left[\sum_{i=1}^{n}\lambda_i\xi_i^2 < c_\alpha\right]$$

$$= \Pr\left[\sum_{i=1}^{ms(T-1)}\lambda_i\xi_i^2 + \sum_{i=ms(T-1)+1}^{m(s-1)}\lambda_i\xi_i^2 + \sum_{i=m(s-1)+1}^{m}\lambda_i\xi_i^2 < c_\alpha\right]$$

$$= \Pr\left[\left\{(1-\rho_{11}-\rho_{22})^{-1}-1\right\}\chi^2_{ms(T-1)} + \left\{(1-\rho_{11}-\rho_{22}(1-T))^{-1}-1\right\}\right.$$

$$\left.\chi^2_{m(s-1)} + \left\{(1+\rho_{11}(sT-1)+\rho_{22}(T-1))^{-1}-1\right\}\chi^2_m < c_\alpha\right], \qquad (3.34)$$

where $\xi=(\xi_1,...,\xi_n)' \sim IN(0, I_n)$ and λ_i, $i=1,...,n$ are the eigenvalues of the matrix $(\Omega^{-1}(\rho_{11}, \rho_{22})-I_n)$, namely $\{(1-\rho_{11}-\rho_{22})^{-1}-1\}$, $\{(1-\rho_{11}-\rho_{22}(1-T)^{-1})-1\}$ and $\{(1+\rho_{11}(sT-1) + \rho_{22}(T-1))^{-1}-1\}$ with multiplicities of ms(T-1), m(s-1) and m, respectively. In order to find c_α such that (3.34) equals the required significance level, α, (3.34) can be evaluated by standard numerical algorithms mentioned in earlier sections.

To calculate the power of the $r(\rho_{11}, \rho_{22})$ test, there is a need to decompose the error covariance matrix $\Omega(\rho_1, \rho_2)$ that is,

$$\Omega(\rho_1,\rho_2) = \Omega^{1/2}(\rho_1,\rho_2)\left[\Omega^{1/2}(\rho_1,\rho_2)\right]',$$

where

$$\Omega^{1/2}(\rho_1,\rho_2) = \bigoplus_{i=1}^{m}\Omega_i^{1/2}(\rho_1,\rho_2),$$

such that the elements of the diagonal components matrix, $\Omega^{1/2}(\rho_1,\rho_2)$ are given in appendix 3B and the formula for the $\Omega_i^{1/2}(\rho_1,\rho_2)$ is:

$$\Omega_i^{1/2}(\rho_1,\rho_2) = \sqrt{\lambda_1}\,\frac{e_{N_i}\tau e'\,_{N_i}\tau}{N_i\tau} + \sqrt{\lambda_2}\left(I_{N_i} \otimes I_\tau - I_{N_i} \otimes \frac{e_\tau e'_\tau}{\tau}\right)$$

$$+ \sqrt{\lambda_3}\left(I_{N_i} \otimes \frac{e_\tau e'_\tau}{\tau} - \frac{e_{N_i}\tau e'\,_{N_i}\tau}{N_i\tau}\right)$$

where

$$\lambda_1 = 1-\rho_1-\rho_2+N_i\tau\rho_1+\tau\rho_2,$$

$$\lambda_2 = 1-\rho_1-\rho_2,$$

$$\lambda_3 = 1-\rho_1-\rho_2+\tau\rho_2. \tag{3.35}$$

For any critical value c_α, the power of the critical region $r(\rho_{11},\rho_{22}) < c_\alpha$ is obtained by evaluating the probabilities of the form,

$$\Pr\left[r(\rho_{11},\rho_{22}) < c_\alpha\right]$$

$$= \Pr\left[Y'\left(\Omega^{-1}(\rho_{11},\rho_{22})-I_n\right)Y < c_\alpha | Y \cdot N\left(0,\Omega(\rho_1,\rho_2)\right)\right]$$

$$= \Pr\left[z'\left(\Omega^{1/2}\right)'\left(\Omega^{-1}(\rho_{11},\rho_{22})-I_n\right)\Omega^{1/2}z < c_\alpha | z \sim N(0,I_n)\right]$$

$$= \Pr\left[\sum_{l=1}^{n}\lambda_i\xi_i^2 < c_\alpha\right]$$

$$= \Pr\left[a\chi^2_{ms(\tau-1)} + b\chi^2_{m(s-1)} + c\chi^2_m < c_\alpha\right], \tag{3.36}$$

where $\xi_i \sim IN(0,1), \chi_j^2$ denotes the usual chi-square distribution with j degrees of freedom and λ_i's are the eigenvalues of

$$\left(\Omega^{1/2}\right)'\left(\Omega^{-1}(\rho_{11},\rho_{22})-I_n\right)\Omega^{1/2},$$

or equivalently, the eigenvalues of

$$\left(\Omega^{-1}(\rho_{11},\rho_{22})-I_n\right)\Omega(\rho_1,\rho_2)$$

$$= \Omega^{-1}(\rho_{11}, \rho_{22}) \Omega(\rho_1, \rho_2) - \Omega(\rho_1, \rho_2),$$

where $\Omega^{1/2} = \Omega^{1/2}(\rho_1, \rho_2)$.

It follows from lemma two together with lemma three of appendix 3A that the eigenvalues of the latter matrix are

$$a = \frac{(1 - \rho_1 - \rho_2)}{(1 - \rho_{11} - \rho_{22})} \, [\rho_{11} + \rho_{22}] \tag{3.37}$$

$$b = \frac{(1 - \rho_1 - \rho_2(1 - T))}{(1 - \rho_{11} - \rho_{22}(1 - T))} \{\rho_{11} + \rho_{22}(1 - T)\}, \tag{3.38}$$

and

$$c = \frac{(1 - \rho_1(1 - sT) - \rho_2(1 - T))}{(1 - \rho_{11}(1 - sT) - \rho_{22}(1 - T))} \{\rho_{11}(1 - sT) + \rho_{22}(1 - T)\}, \tag{3.39}$$

with multiplicities of $ms(T-1)$, $m(s-1)$ and m, respectively.

In the special cases in which power is being calculated at

$$(\rho_1, \rho_2)' = (\rho_{11}, \rho_{22})',$$

then (3.37), (3.38) and (3.39) reduce to

$$\left(\rho_{11} + \rho_{22} \right), \ \left(\rho_{11} + \rho_{22}(1 - T) \right) \text{ and } \left\{ \rho_{11}(1 - sT) + (1 - T) \right\},$$

respectively, so that (3.36) can be written as,

$$\Pr\left[\chi^2_{ms(T-1)} + \left(1 - \frac{T\rho_{22}}{\rho_{11} + \rho_{22}} \right) \chi^2_{m(s-1)} + \right.$$

$$\left. \left(1 - \frac{T(s\rho_{11} + \rho_{22})}{\rho_{11} + \rho_{22}} \right) \chi^2_m < \frac{c_\alpha}{\rho_{11} + \rho_{22}} \right]. \tag{3.40}$$

Under these special circumstances, one can calculate the power of the $r(\rho_1, \rho_2)$ test by solving (3.40). In the next section the PO test for testing subblock effects in the presence of main block effects will be constructed.

PO test for testing subblock effects

In the previous section the PO test for testing whether intrablock and intrasubblock equicorrelation coefficients are equal to zero against the alternative that they have some positive value was derived. The main focus of this section would be on testing the intrasubblock equicorrelation coefficient in the presence of intrablock correlation. Detailed application of such a testing problem in the case of 3SLR model is given in chapter six of this book.

Here in this section the main objective is to test

$$H_0: \ \rho_2 = 0, \ \rho_1 > 0 \tag{3.41}$$

against the one-sided alternative

$$H_a: \rho_2 > 0, \ \rho_1 > 0. \tag{3.42}$$

This problem is more complex than the testing problem of the point optimal test in the previous section. In this case both the null and the alternative hypotheses are composite. However, under H_0 the density function of Y is,

$$f_0(Y:\rho_1,0) = (2\pi)^{-n/2} |\Omega_0|^{-1/2} \exp\left[-\frac{1}{2} Y'\Omega_0^{-1}Y\right]$$

while under H_a it is given by

$$f_a(Y:\rho_1,\rho_2) = (2\pi)^{-n/2} |\Omega_1|^{-1/2} \exp\left[-\frac{1}{2} Y'\Omega_1^{-1}Y\right]$$

where $\Omega_0 = \Omega(\rho_1,0)$ and $\Omega_1 = \Omega(\rho_1,\rho_2)$

Note that $\Omega(\rho_1,\rho_2)$ is defined by (3.26), where $0 \le \rho_1 + \rho_2 \le 1$ and ρ_1 and ρ_2 are unknown parameters. For the purpose of test construction, one can have a couple of options to decide about the range of ρ_1 values. The first option is $0 \le \rho_1 \le 1$, which seems unlikely because $0 \le \rho_1 + \rho_2 \le 1$. Another that might be reasonable to assume is $0 \le \rho_1 \le 0.5$. First of all one can consider the simple problem of testing,

$$H_0'':(\rho_1,\rho_2)' = (\rho_{10},0)',$$

against the simple alternative hypothesis

$$H_a'':(\rho_1,\rho_2)' = (\rho_{11},\rho_{21})',$$

where $0 \le \rho_{10} \le 0.5$, $0 \le \rho_{11} \le 0.5$ and $0 \le \rho_{21} \le 0.5$ are known fixed values. The Neyman-Pearson Lemma implies that a most powerful test can be based on the critical region of the form

$$r(\rho_{10},\rho_{11},\rho_{21}) = Y'\left(\Omega_{11}^{-1} - \Omega_{10}^{-1}\right)Y < c_\alpha, \tag{3.43}$$

where c_α is an appropriate critical value.

Note that $r(\rho_{10},\rho_{11},\rho_{21})$ in (3.43) is a quadratic form in normal variables. Therefore its critical values, c_α, for the required level of significance, α, can be obtained by evaluating the probabilities of the form,

$$= \Pr\left[r(\rho_{10},\rho_{11},\rho_{21}) < c_\alpha \middle| Y \sim N\left(0,\Omega(\rho_{10},0)\right)\right]$$

$$= \Pr\left[Y'\left(\Omega_{11}^{-1} - \Omega_{10}^{-1}\right)Y < c_\alpha \middle| Y \sim N\left(0,\Omega_0\right)\right]$$

$$= \Pr\left[\left(\Omega_0^{-1/2}Y\right)'\left(\Omega_0^{1/2}\right)'\left(\Omega_{11}^{-1} - \Omega_{10}^{-1}\right)\Omega_0^{-1/2}\left(\Omega_0^{-1/2}Y\right) < c_\alpha \middle| Y \sim N\left(0,\Omega_0\right)\right]$$

$$= \Pr\left[z'\left(\Omega_0^{1/2}\right)'\left(\Omega_{11}^{-1} - \Omega_{10}^{-1}\right)\Omega_0^{1/2}z < c_\alpha \middle| z \sim N\left(0,I_n\right)\right]$$

$$= \Pr\left[\sum_{i=1}^{m}\lambda_i\xi_i^2 < c_\alpha\right], \tag{3.44}$$

where $\xi_i \sim IN(0,1)$, and λ_i's are the eigenvalues of
$\left[\left(\Omega_0^{1/2}\right)'\left(\Omega_{11}^{-1} - \Omega_{10}^{-1}\right)\Omega_0^{1/2}\right]$ or equivalently, the eigenvalues of

$$\left[\left(\Omega_{11}^{-1} - \Omega_{10}^{-1}\right)\Omega_0\right] \tag{3.45}$$

in which $\Omega_{11} = \Omega(\rho_{11},\rho_{22})$, $\Omega_{10} = \Omega(\rho_{10},0)$ and $\Omega_0 = \Omega(\rho_1,0)$.

Using lemmas three and four of appendix 3A, together with lemma two of the same appendix, the eigenvalues of the latter matrix can be found, as follows,

$$a_1 = (1-\rho_1)\left\{\frac{1}{1-\rho_{11}-\rho_{22}} - \frac{1}{1-\rho_{10}}\right\}, \tag{3.46}$$

$$b_1 = \frac{(1-\rho_1)}{1-\rho_{11}-\rho_{22}}\left\{\frac{1-\rho_{11}-\rho_{22}}{1-\rho_{11}-\rho_{22}+T\rho_{22}} - \frac{1}{1-\rho_{10}}\right\}, \tag{3.47}$$

and

$$c_1 = (1-\rho_1+sT\rho_1)\left[\frac{1}{(1-\rho_{11}-\rho_{22}+T\rho_{22}+sT\rho_{11})}\right.$$
$$\left. - \left(\frac{1}{(1-\rho_{10})(1-\rho_1+sT\rho_1)} - \frac{\rho_{10}}{(1-\rho_{10})(1-\rho_{10}+sT\rho_{10})}\right)\right], \tag{3.48}$$

with multiplicities $ms(T-1)$, $m(s-1)$ and m, respectively.

Therefore, one can write (3.43) as,

$$\Pr\left[a_1\chi^2_{ms(T-1)} + b_1\chi^2_{m(s-1)} + c_1\chi^2_m < c_\alpha\right], \tag{3.49}$$

where a_1, b_1 and c_1 are given by (3.46), (3.47) and (3.48), respectively.

Thus a test based on $r(\rho_{10},\rho_{11},\rho_{21})$ in (3.43), is most powerful in the neighbourhood of $(\rho_1,\rho_2)' = (\rho_{11},\rho_{21})'$. Therefore it is a PO test of a simple null $H_0^{''}$ against a simple alternative $H_a^{''}$. For the wider problem of testing (3.41) against (3.42) it may not necessarily be PO, because for this testing problem the critical value should be found by solving

$$\sup_{0<\rho_1\leq 0.5} \Pr\left[r(\rho_{10},\rho_{11},\rho_{21}) < c_\alpha^*\big|y \sim N\left(0,\Omega(\rho_1,0)\right)\right] = \alpha \tag{3.50}$$

for c_α^*.

In general, $c_\alpha < c_\alpha^*$ so the two critical regions are different. As the range of ρ_1 is closed

$$\Pr\left[r(\rho_{10},\rho_{11},\rho_{21}) < c_\alpha^* \middle| y \sim N(0,\Omega(\rho_1,0))\right] = \alpha$$

is true for at least one ρ_1 value. If one can find such a ρ_{10} to let $c_\alpha = c_\alpha^*$ can be found, then the test based on $r(\rho_{10},\rho_{11},\rho_{21})$ is MP and is a PO test of (3.41) against (3.42). However, in general, such a ρ_{10} value may not exist. Perhaps it may exist for some combinations of ρ_{11}, ρ_{21}, m, s and T values, but not for others. If the value of the ρ_{10} does exist (as will be seen in chapter six dealing with the 3SLR model), then (3.46), (3.47) and (3.48) become

$$a_1' = \left\{ \frac{1-\rho_{10}}{1-\rho_{11}-\rho_{22}} - 1 \right\},$$

$$b_1' = \frac{1}{1-\rho_{11}-\rho_{22}} \left\{ \frac{(1-\rho_{11}-\rho_{22})(1-\rho_{10})}{(1-\rho_{11}-\rho_{22}+T\rho_{22})} - 1 \right\}$$

and

$$c_1' = \left\{ \frac{1-\rho_{10}+sT\rho_{10}}{(1-\rho_{11}-\rho_{22}+T\rho_{22}+sT\rho_{11})} - 1 \right\},$$

respectively. Therefore c_α can be obtained by solving the left hand side of,

$$\Pr\left[a_1'\chi^2_{ms(T-1)} + b_1'\chi^2_{m(s-1)} + c_1'\chi^2_m < c_\alpha\right] = \alpha,$$

where α is the desired level of significance, and χ^2_j is the usual chi-squared distribution with j degrees of freedom. If in case ρ_{10} does not exist then one can look for an approximate PO test which was discussed in chapter two.

Concluding remarks

This chapter considered an nx1 dimensional random vector, Y, which follows either the two- or the three-stage SSMN distribution and derived some optimal tests for detecting the equicorrelation coefficient between observations within blocks and even within subblocks. It also considered more complicated problems for detecting subblock equicorrelation in the presence of main block equicorrelation.

In the case of the two-stage SSMN distribution model, the PE with the powers of SenGupta's LB test and two versions of the BO tests (ie, $r_{0.5}$ and $r_{0.8}$) at the five percent level of significance for different blocks and sample sizes was computed and compared. Selected one and five percent significance points, c_α, and their associated ρ_1 values are given in tables 3.1 and 3.2 of appendix 3C whereas the power of comparison for the various values of k and m are presented in tables 3.3, 3.4 and 3.5 of the same appendix. The main findings of this chapter are the BO tests are usually superior to that of the LB test, sometimes by a wide margin, while the power differences between the PE and BO tests are very small, the biggest difference being 0.031. On the basis of these results it was argued that the BO tests are approximately UMP tests, at least at the five percent level of significance.

Encouraged by the small sample power performances of BO tests for the two-stage SSMN model, this chapter enabled us to derive POI and LMMPI tests for the three-stage SSMN model. The next logical step is to consider a model based on the SMN distribution (ie, when σ^2 is unknown) and discover the consequences of applying optimal tests to it. This is done in the next chapter.

Notes

1   A paper presenting some of the findings reported in this chapter were published in the Australian Journal of Statistics, Vol. 32(1), p. 87-97, co-authored with Professor Maxwell L. King.

2   This is a special case of the 3SLR model discussed at (2.5).

3   Similar format of ρ_1 and ρ_2 in relation to the 3SLR model is defined by (2.6).

Appendix 3A: Some useful matrix results

This appendix gives some important matrix results which are useful in chapters three, four, five and six.

Lemma 1

Let $\Delta(\rho) = (1-\rho)I_n + \rho D$, where $\dfrac{-1}{n-1} < \rho < 1$, D is given by (3.5) and $n = mk$, then one can have the following expressions.

(i) $\quad |\Delta(\rho)| = \left[(1-\rho)^{k-1}[1+(k-1)]\rho\right]^m$.

(ii) $\quad \Delta^{-1}(\rho) = (1-\rho)^{-1}\left[I_n - \rho\{1+(k-1)\rho\}^{-1}D\right]$.

(iii) The eigenvalues of $\Delta^{-1}(\rho)$ are $\dfrac{1}{(1-\rho)}$ and $\dfrac{1}{[1+(k-1)\rho]}$ with multiplicity m(k-1) and m, respectively.

(iv) Theficance level for $m=10[1+(k-1)\rho]$ with multiplicity m(k-1) and m, respectively.

(v) The matrix $\Delta(\rho)$, can be expressed in functional form as

$$\Delta(\rho) = \left(f_1(\rho)I_n + f_2(\rho)D\right)^{-1}$$

where
$$f_1(\rho) = (1-\rho)^{-1}$$
$$f_2(\rho) = -\rho\left[(1-\rho)\{1+(k-1)\rho\}\right]^{-1}.$$

The proof is straightforward and is therefore omitted. For the special case when m=1, this lemma reduces to Rao (1973, p. 67, problem 1.1)

Lemma 2

If A is nxn and B is nxn, then the eigenvalues of AB are the engenvalues of BA, although the eigenvectors of the two matrices need not be the same (see Graybill, 1969, p. 208).

Lemma 3

Let $D_1 = I_m \otimes E_{sT}$, $D_2 = I_{ms} \otimes E_T$, $n = msT$ and $\lambda_i(A)$ denote the ith eigenvalue of the matrix A, then

$$\lambda_i(D_1 + D_2) = \lambda_i(D_1) + \lambda_i(D_2),$$

for i=1,...,n. These eigenvalues are T(s+1), T and zeroes with multiplicities of m, m(s-1), and ms(T-1), respectively.

Lemma 4

Let $\Omega(\rho_1,\rho_2) = (1-\rho_1-\rho_2)I_n + \rho_1 D_1 + \rho_2 D_2$, where $0 \le \rho_1 + \rho_2 \le 1$, then

(i) The eigenvalues of $\Omega(\rho_1,\rho_2)$ are $(1-\rho_1-\rho_2)$,

$\{1-\rho_1-\rho_2(1-T)\}$ and $\{1+\rho_1(sT-1)+\rho_2(T-1)\}$ with multiplicities ms(T-1), m(s-1) and m, respectively.

(ii) $\Omega^{-1}(\rho_1,\rho_2) = (1-\rho_1-\rho_2)^{-1}\big[I_n - (1-\rho_1+\rho_2(T-1))^{-1}$

$\{\rho_1(1-\rho_1-\rho_2)(1-\rho_1-\rho_2+T(\rho_2+s\rho_1)^{-1}D_1 + \rho_2 D_2\}\big].$

(iii) The eigenvalues of $\Omega^{-1}(\rho_1,\rho_2)$ are the reciprocal of the eigenvalues of $\Omega(\rho_1,\rho_2)$.

The results of lemmas three and four are the special cases which are obtained by using Searle and Henderson's (1974) general procedure.

Appendix 3B: Decomposition of the covariance matrix

Equation (3.35) is proved by following Nerlove's (1971) procedure. Recall that (3.27) is

$$\Omega_i(\rho_1,\rho_2) = (1-\rho_1-\rho_2)\left(I_{N_i} \otimes I_T\right) + \rho_1\left(e_{N_i} e'_{N_i} \otimes e_T e'_T\right)$$

$$+\rho_2\left(I_{N_i} \otimes e_T e'_T\right).$$

Let N_i-1, $N_i \times 1$ vectors ϕ_j, $j = 1,\dots,N_i-1$, satisfy

$$e'_{N_i}\phi_j = 0$$

$$\phi'_j\phi_{j'} = \begin{cases} 1, & j=j' \\ 0, & j\neq j' \end{cases} \quad j,j'=1,\dots,N_i \times 1$$

and $T-1$. $T \times 1$ vectors ψ_k, $k = 1,\dots,T-1$ satisfy

$$e'_T\psi_k = 0$$

$$\psi'_k\psi'_k = \begin{cases} 1, & k=k' \\ 0, & k\neq k' \end{cases} \quad k,k'=1,\dots,T-1$$

Then it is easy to verify that $\Omega_i(\rho_1,\rho_2)$ has three distinct eigenvalues.

These are:

$$\lambda_1 = 1-\rho_1-\rho_2 + N_i T\rho_1 + T\rho_2 \text{ with one associated eigenvector}$$

$$e_{N_i} / \sqrt{N_i} \otimes e_T / \sqrt{T};$$

$$\lambda_2 = 1-\rho_1-\rho_2 \text{ with } (N_i - 1)(T - 1)+T - 1 \text{ associated eigenvectors}$$

$$\phi_j \otimes \psi_k \text{ and } e_{N_i} / \sqrt{N_i} \otimes \psi_k;$$

$$\lambda_3 = 1-\rho_1-\rho_2 + T\rho_2 \text{ with } N_i - 1 \text{ associated eigenvectors}$$

$$\phi_j \otimes e_T / \sqrt{T}.$$

Putting these three sets of eigenvectors into three matrices, results in

$$C_1 = e_{N_i} / \sqrt{N_i} \otimes e_T / \sqrt{T} = e_{N_iT} / \sqrt{N_iT}, \text{ an } N_iT\text{x1 vector,}$$

$$C_2 = \left(\phi_1 \otimes \psi_1, ..., \phi_{N_i-1} \otimes \psi_{T-1}, e_{N_i} / \sqrt{N_i} \otimes \psi_1, ..., e_{N_i} \right.$$
$$\left. / \sqrt{N_i} \otimes \psi_{T-1} \right), \text{ an } N_iT\text{x}N_i(T-1) \text{ matrix,}$$

$$C_3 = \left(\phi_1 \otimes e_T / \sqrt{T}, ..., \phi_{N_i-1} \otimes e_T / \sqrt{T} \right), \text{ an } N_iT\text{x}(N_i\text{-}1) \text{ matrix. So}$$

$$\Omega^{1/2}(\rho_1,\rho_2) = (C_1,C_2,C_3) \begin{pmatrix} \sqrt{\lambda_1} & 0 & 0 \\ 0 & \sqrt{\lambda_2}I_{N_i(T-1)} & 0 \\ 0 & 0 & \sqrt{\lambda_3}I_{N_i-1} \end{pmatrix} \begin{pmatrix} C_1' \\ C_2' \\ C_3' \end{pmatrix}$$

$$= \sqrt{\lambda_1}C_1C_1' + \sqrt{\lambda_2}C_2C_2' + \sqrt{\lambda_3}C_3C_3'.$$

Obviously

$$C_1C_1' - \frac{e_{N_iT}e_{N_iT}'}{N_iT}.$$

To obtain the expressions of C_2C_2' and C_3C_3', it is noted that due to the orthogonality of the eigenvector matrix of $\Omega_i(\rho_1,\rho_2)$

$$CC' = C_1C_1' + C_2C_2' + C_3C_3' = I_{N_iT}$$

where $C = (C_1,C_2,C_3)$. Since

$$\left(I_{N_i} \otimes e_Te_T' \right)C_1C_1' = e_{N_iT}e_{N_iT}' / N_i$$

$$\left(I_{N_i} \otimes e_Te_T' \right)C_2C_2' = 0$$

$$\left(I_{N_i} \otimes e_Te_T' \right)C_3C_3' = TC_3C_3',$$

multiplying CC' by $I_{N_i} \otimes e_Te_T'$ from the left, to get

61

$$C_3 C_3' = I_{N_i} \otimes \frac{e_T e_T'}{T} - \frac{e_{N_i T} e_{N_i T}'}{N_i T}$$

and

$$C_2 C_2' = CC' - C_1 C_1' - C_3 C_3'$$

$$= I_{N_i T} - I_{N_i} \otimes \frac{e_T e_T'}{T}.$$

Appendix 3C: Tabulation of results of the comparative study

Table 3.1

Selected values of ρ_1 and c_α for the $r_{0.5}$ test at the one and five percent significance levels

k	α	$m=5$		$m=10$		$m=15$		$m=25$	
		ρ_1	c_α	ρ_1	c_α	ρ_1	c_α	ρ_1	c_α
2	0.01	.6694	10^{-4}	.6581	10^{-5}	.5577	10^{-5}	.4451	10^{-5}
	0.05	.8329	-10^{-5}	.4973	-10^{-5}	.4124	-10^{-6}	.3233	10^{-5}
3	0.01	.6269	.3779	.4531	.2885	.3700	.2394	.2850	.1867
	0.05	.4610	.2778	.3227	.2055	.2012	.1690	.2002	.1311
4	0.01	.4996	.6146	.3461	.4448	.2779	.3618	.2107	.2770
	0.05	.3530	.4342	.2403	.3088	.1924	.2504	.1459	.1918
5	0.01	.4151	7748	.2803	.5430	.2228	.4364	.1674	.3308
	0.05	.2864	.5346	.1917	.3714	.1525	.2988	.1149	.2271
6	0.01	.3550	.8900	.2356	.6105	.1860	.4869	.1389	.3664
	0.05	.2412	.6046	.1596	.4136	.1264	.3309	.0949	.2503
7	0.01	.3102	.9769	.2032	.6598	.1597	.5233	.1187	.3918
	0.05	.2084	.6562	.1367	.4440	.1079	.3538	.0808	.2667
8	0.01	.2754	1.045	.1787	.6973	.1399	.5507	.1037	.4109
	0.05	.1835	.6960	.1196	.4670	.0942	.3709	.0704	.2792
9	0.01	.2477	1.099	.1594	.7270	.1245	.5722	.0920	.4257
	0.05	.1639	.7275	.1063	.4849	.0836	.3843	.0623	.2884
10	0.01	.2250	1.442	.1439	.7509	.1121	.5895	.0828	.4378
	0.05	.1481	.7531	.0957	.493	.0751	.3949	.0560	.2959

Table 3.2
Selected values of ρ_1 and c_α for the $r_{0.8}$ test at the one and five percent significance levels

k	α	$m=5$ ρ_1	c_α	$m=10$ ρ_1	c_α	$m=15$ ρ_1	c_α	$m=25$ ρ_1	c_α
2	0.01	.9114	3.068	.7767	3.906	.6835	4.289	.5656	4.653
	0.05	.8189	2.756	.6624	3.331	.5697	3.575	.4615	3.796
3	0.01	.7657	4.680	.5942	5.338	.4980	5.546	.3910	5.672
	0.05	.6469	3.953	.4821	4.331	.3979	4.431	.3086	4.476
4	0.01	.6584	5.784	.4833	6.212	.3940	6.263	.3008	6.211
	0.05	.5384	4.730	.3823	4.914	.3082	4.900	.2336	4.824
5	0.01	.5788	6.621	.4084	6.819	.3268	6.739	.2448	6.552
	0.05	.4633	5.299	.3178	5.306	.2523	5.203	.1883	5.039
6	0.01	.5173	7.289	.3541	7.270	.2794	7.080	.2066	6.788
	0.05	.4077	5.744	.2724	5.592	.2138	5.416	.1579	5.187
7	0.01	.4684	7.840	.3128	7.621	.2442	7.337	.1788	6.962
	0.05	.3646	6.103	.2385	5.811	.1856	5.576	.1360	5.294
8	0.01	.4284	8.305	.2803	7.901	.2170	7.539	.1577	7.095
	0.05	.3302	6.401	.2123	5.983	.1640	5.699	.1195	5.376
9	0.01	.3950	8.705	.2540	8.132	.1952	7.701	.1410	7.201
	0.05	.3019	6.654	.1913	6.124	.1470	5.797	.1065	5.441
10	0.01	.3666	9.052	.2323	8.324	.1775	7.835	.1275	7.287
	0.05	.2783	6.871	.1741	6.240	.1331	5.878	.0961	5.493

Table 3.3
Comparison of the PE with the powers of LB, $r_{0.5}$ and $r_{0.8}$ tests at the five percent significance level for m=10

Tests

	$\rho=.05$	.1	.2	.3	.4	.5	.6	.8
				k=2				
PE	.069	.093	.155	.242	.357	.504	.683	.983
LB	.069	.092	.153	.213	.325	.430	.538	.737
$r_{0.5}$	.067	.088	.147	.234	.354	.504	.675	.952
$r_{0.8}$	.065	.083	.135	.215	.331	.489	.680	.978
				k=3				
PE	.088	.139	.279	.456	.650	.829	.951	1.000
LB	.088	.139	.270	.423	.572	.701	.802	.926
$r_{0.5}$	.085	.135	.275	.456	.646	.809	.920	1.000
$r_{0.8}$	.081	.125	.255	.438	.646	.829	.946	1.000
				k=4				
PE	.109	.192	.406	.636	.830	.950	.994	1.000
LB	.109	.190	.389	.582	.735	.841	.909	.974
$r_{0.5}$	.106	.188	.405	.633	.812	.922	.975	.999
$r_{0.8}$	.100	.174	.387	.631	.830	.945	.990	1.000
				k=6				
PE	.155	.305	.622	.851	.964	.997	1.000	1.000
LB	.155	.301	.589	.785	.893	.949	.976	.995
$r_{0.5}$	.153	.303	.621	.836	.941	.983	.996	1.000
$r_{0.8}$	.144	.288	.617	.851	.958	.992	.999	1.000
				k=10				
PE	.261	.521	.861	.977	.999	1.000	1.000	1.000
LB	.259	.509	.820	.936	.977	.991	.997	.999
$r_{0.5}$	.259	.521	.849	.960	.990	.998	1.000	1.000
$r_{0.8}$	.249	.513	.860	.971	.996	1.000	1.000	1.000

Table 3.4
Comparison of the PE with the powers of LB, $r_{0.5}$ and $r_{0.8}$ tests at the five percent significance level for m=15

Tests

	$\rho=.05$	.1	.2	.3	.4	.5	.6	.8
				$k=2$				
PE	.074	.105	.191	.315	.478	.668	.852	.999
LB	.074	.104	.188	.299	.431	.569	.699	.889
$r_{0.5}$	.072	.101	.186	.312	.478	.663	.833	.993
$r_{0.8}$	.070	.096	.174	.295	.464	.665	.851	.998
				$k=3$				
PE	.098	.167	.359	.591	.804	.943	.994	1.000
LB	.098	.165	.346	.547	.720	.844	.922	.985
$r_{0.5}$	.096	.163	.358	.590	.792	.921	.979	1.000
$r_{0.8}$	.092	.154	.343	.585	.804	.939	.990	1.000
				$k=4$				
PE	.125	.237	.522	.783	.939	.992	1.000	1.000
LB	.125	.235	.499	.724	.867	.942	.977	.997
$r_{0.5}$	.123	.235	.522	.774	.920	.979	.996	1.000
$r_{0.8}$	.118	.224	.514	.783	.935	.988	.999	1.000
				$k=6$				
PE	.186	.388	.763	.947	.995	1.000	1.000	1.000
LB	.186	.381	.726	.901	.967	.990	.997	1.000
$r_{0.5}$	.185	.388	.757	.932	.986	.998	1.000	1.000
$r_{0.8}$	.178	.378	.763	.944	.992	.999	1.000	1.000
				$k=10$				
PE	.327	.650	.950	.997	1.000	1.000	1.000	1.000
LB	.325	.635	.922	.985	.997	.999	1.000	1.000
$r_{0.5}$	.326	.649	.940	.992	.999	1.000	1.000	1.000
$r_{0.8}$	.320	.649	.947	.995	1.000	1.000	1.000	1.000

Table 3.5
Comparison of the PE with the powers of LB, $r_{0.5}$ and $r_{0.8}$ tests at the five percent significance level for m=25

Tests	$\rho=.05$	.1	.2	.3	.4	.5	.6	.8
				k=2				
PE	.082	.126	.259	.449	.672	.868	.975	1.000
LB	.082	.126	.253	.424	.609	.773	.889	.986
$r_{0.5}$	.080	.123	.256	.449	.669	.854	.960	1.000
$r_{0.8}$	.078	.119	.246	.440	.670	.867	.972	1.000
				k=3				
PE	.115	.216	.499	.780	.946	.995	1.000	1.000
LB	.115	.215	.480	.731	.889	.964	.990	1.000
$r_{0.5}$	.114	.214	.499	.774	.932	.988	.999	1.000
$r_{0.8}$	.110	.207	.493	.780	.943	.993	1.000	1.000
				k=4				
PE	.154	.320	.700	.930	.994	1.000	1.000	1.000
LB	.153	.316	.670	.887	.970	.994	.999	1.000
$r_{0.5}$	.152	.319	.697	.919	.987	.999	1.000	1.000
$r_{0.8}$	.148	.311	.699	.928	.991	.999	1.000	1.000
				k=6				
PE	.243	.529	.912	.994	1.000	1.000	1.000	1.000
LB	.242	.519	.884	.981	.997	1.000.	1.000	1.000
$r_{0.5}$	.242	.529	.905	.989	.999	1.000	1.000	1.000
$r_{0.8}$	.237	.526	.911	.992	1.000	1.000	1.000	1.000
				k=10				
PE	.443	.820	.994	1.000	1.000	1.000	1.000	1.000
LB	.440	.805	.987	.999	1.000	1.000	1.000	1.000
$r_{0.5}$	.443	.817	.991	1.000	1.000	1.000	1.000	1.000
$r_{0.8}$	.440	.820	.993	1.000	1.000	1.000	1.000	1.000

4 Some tests of the symmetric multivariate normal distribution

Introduction

In chapter three it was noted that the LB, BO, PO and LMMP tests have certain small sample optimal power properties for the cases of two- and three-stage SSMN distribution models. The main objective of this chapter is to consider analogous tests for the case of SMN distribution models in which σ^2 is an unknown scalar parameter. This chapter generalizes the results of the previous chapter in constructing the PO invariant (POI) and LMMP invariant (LMMPI) tests. It also demonstrates that the POI test for the two-stage SMN model is UMP invariant (UMPI). The most attractive features of the POI test in this case is that its power and critical values are obtainable from the standard F distribution. Among other things this chapter also explores the possibility of developing some optimal tests based on the three-stage SMN distribution model.

The structure of this chapter is as follows. In the next section, the two-stage SMN model is introduced and the POI test is constructed. It shows how one can calculate the power and the critical values of the POI test with the help of the standard F distribution and provides a proof of the fact that the test is UMPI. It also constructs a LMMPI test for a more complicated model, where means are known but variances and the equicorrelation coefficient are unknown. The following section demonstrates how to derive LMMPI and POI tests for the three-stage SMN model and develops the POI test in testing subblock effects in the presence of main block effects. The final section contains some concluding remarks.

Two-stage SMN model and the tests

So far testing problems associated with the SSMN model have been considered. Although the definition and examples of the model which follow the SMN distribution was discussed in chapter two but no test for testing equicorrelation coefficient based on SMN model was developed in the previous chapter. The main objective of this section is to construct some optimal tests based on the two-stage SMN model. The model under consideration is,

$$Y \sim N\left(0, \sigma^2 \Delta(\rho)\right),$$
(4.1)

where $\Delta(\rho)$ is given by (3.4) via (3.1) or equivalently (2.3). Additionally, throughout this chapter a more practical assumption that σ^2 is an unknown scalar parameter is made. Based on the model (4.1), one can easily construct the point optimal test which is given in the next sub-section.

Point optimal invariant test

Following the previous chapter, the problem of interest in this subsection is to test

$$H_0: \rho = 0$$

against

$$H_a: \rho > 0.$$

Note that this testing problem is invariant (in the sense of Lehmann, 1986 and of the discussion in chapter two) under the group of transformations

$$Y \to \eta_0 Y$$
(4.2)

where η_0 is a positive scalar. The vector

$$\omega = Y / (Y'Y)^{1/2}$$
(4.3)

is maximal invariant under this group of transformations. The joint density of ω would be

70

$$f(\omega) = \frac{1}{2}\Gamma\left(\frac{n}{2}\right)\pi^{-n/2}|\Delta(\rho)| - \frac{1}{2}\left(\omega'\Delta^{-1}(\rho)\omega\right)^{-n/2}$$

(4.4)

with respect to the uniform measure on

$$\{\omega : \omega \in \mathbb{R}^n, \omega'\omega = 1\}.$$

Hence the joint density function of ω, under H_a is (4.4), while under H_0 it reduces to

$$f_0(\omega) = \frac{1}{2}\Gamma\left(\frac{n}{2}\right)\pi^{-n/2}.$$

To begin with, let us consider the simpler problem of testing H_0 against the simple alternative hypothesis $H_a : \rho = \rho_1 > 0$, in (4.1), with ρ_1 a known and fixed point. The Neyman-Pearson Lemma yields the critical region of the form

$$|\Delta(\rho_1)|^{-1/2}\left(\omega'\Delta^{-1}(\rho_1)\omega\right)^{-n/2} \geq c$$

or equivalently,

$$s(\rho_1) = \frac{Y'\Delta^{-1}(\rho_1)Y}{Y'Y} < c_\alpha$$

(4.5)

which is the most powerful test, where c_α is an appropriate critical value. For the wider problem of testing H_0 against H_a, the test based on $s(\rho_1)$ in (4.5), is the most powerful invariant at $\rho = \rho_1$ and is therefore called a POI test. The α level critical values can be found by solving the probabilities of the form,

$$\Pr\left[s(\rho_1) < c_\alpha\right]$$

$$= \Pr\left[Y'\left(\Delta^{-1}(\rho_1) - c_\alpha I_n\right)Y < 0\right]$$

$$= \Pr\left[\sum_{i-1}^{n}\lambda_i\xi_i^2 < 0\right]$$

$$= \Pr\left[\left(\frac{1}{1-\rho_1} - c_\alpha\right)\chi^2_{m(k-1)} + \left(\frac{1}{1+(k-1)\rho_1} - c_\alpha\right)\chi^2_m < 0\right], \qquad (4.6)$$

where $\lambda_i, i=1,...,n$, are the eigenvalues of the matrix $\left(\Delta^{-1}(\rho_1) - c_\alpha I_n\right)$, namely $\left(\frac{1}{1-\rho_1} - c_\alpha\right)$ and $\left[\frac{1}{\{1+(k-1)\rho_1\}} - c_\alpha\right]$ with multiplicities of $m(k-1)$ and m respectively, and $\xi_i \sim IN(0, 1)$.

Alternatively, one can note that the critical values of the $s(\rho_1)$ test can also be obtained by the standard F distribution. Notice that equation (4.6) can be expressed as

$$\Pr\left[s(\rho_1) < c_\alpha\right]$$

$$= \Pr\left[(a - c_\alpha)\chi^2_{m(k-1)} + (b - c_\alpha)\chi^2_m < 0\right]$$

$$= \Pr\left[(a - c_\alpha)\chi^2_{m(k-1)} < (c_\alpha - b)\chi^2_m\right]$$

$$= \Pr\left[\frac{\chi^2_{m(k-1)}/m(k-1)}{\chi^2_m / m} < \frac{(c_\alpha - b)}{(a - c_\alpha)(k-1)}\right]$$

$$= \Pr\left[F_{(m(k-1),m)} < F_\alpha\right],$$

where

$$F_\alpha = \frac{(c_\alpha - b)}{(a - c_\alpha)(k-1)},$$

$$a = \frac{1}{1-\rho_1} \quad \text{and} \quad b = \frac{1}{\{1+(k-1)\rho_1\}},$$

and $F_{(m(k-1),m)}$ is the standard F distribution with $m(k-1)$ and m degrees of freedom. If F_α is the $100(1-\alpha)$ percentile of the F distribution (obtainable from the standard tables), then the critical values c_α can be calculated by the following expression:

$$c_\alpha = \frac{\{b + a(k-1)F_\alpha\}}{\{1 + (k-1)F_\alpha\}}.$$

Thus for any critical value c_α, the power of the $s(\rho_1)$ test is obtained by solving probabilities of the form,

$$\Pr\left[s(\rho_1) < c_\alpha \middle| Y \sim N(0, \sigma^2 \Delta(\rho))\right]$$

$$= \Pr\left[Y'\left(\Delta^{-1}(\rho_1) - c_\alpha I_n\right)Y < 0 \middle| H_a\right]$$

$$= \Pr\left[z'T'\left(\Delta^{-1}(\rho_1) - c_\alpha I_n\right)Tz < 0\right]$$

$$= \Pr\left[\sum_{i=1}^{n} \lambda_i \xi_i^2 < 0\right], \tag{4.7}$$

where λ_i, $i = 1, 2, ..., n$, are the eigenvalues of

$$T'\left(\Delta^{-1}(\rho_1) - c_\alpha I_n\right)T \text{ and } \xi = (\xi_1, ..., \xi_n)' \sim IN(0, I_n).$$

The second last equality of (4.7) follows from the Cholesky decomposition of $\Delta(\rho) = TT'$, where the matrix T is obtained by the scheme given by (3.9) of chapter three. Both (4.6) and (4.7) can be evaluated by standard numerical algorithms mentioned earlier.

As in chapter three, the results of appendix 3A can be used to obtain the eigenvalues of the matrix $T'\left(\Delta^{-1}(\rho_1) - c_\alpha I_n\right)T$, in (4.7), analytically. Hence, note that the eigenvalues of $T'\left(\Delta^{-1}(\rho_1) - c_\alpha I_n\right)T$ are also the eigenvalues of $\left(\Delta^{-1}(\rho_1) - c_\alpha I_n\right)\Delta(\rho)$, namely

$$= \left(\frac{1}{1-\rho_1} - c_\alpha\right)(1-\rho)I_n + \left[\left(\frac{1}{1-\rho_1} - c_\alpha\right)\rho - \frac{\rho_1(1-\rho) + \rho_1\rho k}{\{1+\rho_1(k-1)\}(1-\rho_1)}\right]D$$

$$= a * I_n + b * D$$

say. These eigenvalues are

$$a* = (1-\rho)\left(\frac{1}{1-\rho_1} - c_\alpha\right)$$

(4.8)

and

$$a*+b*k = [1+\rho(k-1)]\left[\frac{1}{1+\rho_1(k-1)} - c_\alpha\right]$$

(4.9)

with multiplicities of m(k - 1) and m, respectively.

Thus one can write (4.7) as,

$$Pr\left[a*\chi^2_{m(k-1)} + (a*+b*k)\chi^2_m < 0\right]$$

$$= Pr\left[a*\chi^2_{m(k-1)} < -(a*+b*k)\chi^2_m\right]$$

$$= Pr\left[\frac{\chi^2_{m(k-1)}/m(k-1)}{\chi^2_m/m} < \frac{-(a*+b*k)}{a*(k-1)}\right]$$

$$= Pr\left[F_{(m(k-1),m)} < F_p\right]$$

where p denotes the power and F_p is obtained from

$$F_p = \frac{-(a*+b*k)}{a*(k-1)},$$

where a* and (a* + b*k) are given by (4.8) and (4.9), respectively.

As is noticed above, one of the features of the $s(\rho_1)$ test is that its power and critical values are obtainable from the standard F distribution. Secondly, note that the $s(\rho_1)$ test is UMPI test as proved below.

Theorem 4.1

The $s(\rho_1)$ test is a UMPI test.

Proof:

74

Consider equation (4.5), i.e.,

$$s(\rho_1) = \frac{Y' \Delta^{-1}(\rho_1) Y}{Y'Y} < c_\alpha$$

where $\Delta^{-1}(\rho_1)$ in (3.6) can be written as,

$$\Delta^{-1}(\rho_1) = aI_n + bD,$$

with

$$a = \frac{1}{1-\rho_1},$$

$$b = \frac{-\rho_1}{(1-\rho_1)\{1+(k-1)\rho_1\}}, \quad \text{then}$$

$$s(\rho_1) = \frac{Y'(aI_n + bD)Y}{Y'Y} < c_\alpha$$

$$s(\rho_1) = a + b\frac{Y'DY}{Y'Y} < c_a$$

or equivalently

$$\frac{Y'DY}{Y'Y} > d_\alpha,$$

where

$$d_\alpha = \frac{c_\alpha - a}{b}.$$

Note that in the above inequality b is negative therefore when it is multiplied by 1/b the inequality reverses. Expressed in this form the critical value d_α is obtained by reference to the null distribution of $\frac{Y'DY}{Y'Y}$. Hence the critical region of the $s(\rho_1)$ test can be written in a way that is independent of ρ_1, and therefore, the test is UMPI.

A more complicated model and the LMMPI test

Consider an nx1 dimensional random vector Y such that

$$Y \sim N\left(0, \sigma^2 \Delta(\rho)\right),$$

where $\Delta(\rho)$ is given as (3.15), with the addition of σ^2 as an unknown scalar and $\rho = (\rho_1, ..., \rho_m)'$. Here the problem of interest is to test the null hypothesis $H_0 : \rho_1 = ... = \rho_m = 0$, against $H_a : \rho_i \geq 0$, for i=1,2,...,m. Note that this testing problem is invariant under the group of transformations (4.2), with the maximal invariant (4.3) and the density of the maximal invariant which is already given by (4.4). Let

$$A_i = -\frac{\partial \Delta(\rho)}{\partial \rho_i}\Bigg|_{\rho=0} = \frac{\partial \Delta^{-1}(\rho)}{\partial \rho_i}\Bigg|_{\rho=0},$$

where $\Delta^{-1}(\rho)$ is given by (3.18), and the elements of the matrix A_i are given by (3.21).

By using equation (3) of King and Wu's (1990) theorem and the density (4.4), the LMMPI test for this model is based on the critical region of the form

$$\sum_{i=1}^{m} -\frac{1}{2}\frac{\partial \ell n |\Delta(\rho)|}{\partial \rho_i}\Bigg|_{\rho=0} - \sum_{i=1}^{m}\frac{n}{2}\frac{\partial \ell n\left(\omega' \Delta^{-1}(\rho)\omega\right)}{\partial \rho_i}\Bigg|_{\rho=0} > c.$$

The first summation on the left is simply a scalar constant. Evaluating the second summation gives the critical region

$$-\frac{1}{2}c_1 - \frac{n}{2}\sum_{i=1}^{m}\omega' A_i \omega > c$$

or equivalently

$$s = \sum_{i=1}^{m} s_i = \sum_{i=1}^{m}\frac{Y' A_i Y}{Y' Y} = \frac{Y' A Y}{Y' Y}$$

or equivalently,

$$s = \frac{Y'DY}{Y'Y} > (1 - c_2), \tag{4.10}$$

where $A = (I_n - D)$ and c_2 is a suitably chosen constant. Note that (4.10) is also the UMPI test of testing H_0 against H_a. Further, note that (4.10) can also be simplified as

$$\frac{\sum\limits_{i=1}^{m}\left(\sum\limits_{j=1}^{k} y_{ij}\right)^2}{\sum\limits_{i=1}^{m}\sum\limits_{j=1}^{k} y_{ij}^2} > (1 - c_2)$$

and its critical values and the power can also be obtained via standard F distribution in a similar way to that of the $s(\rho_1)$ test in the previous section.

Three-stage SMN model and the tests

This section extends the two-stage model to the situation in which observations are obtained from a three-stage block design, i.e.,

$$Y \sim N\left(0, \sigma^2 \Omega(\rho_1, \rho_2)\right), \tag{4.11}$$

where $\Omega(\rho_1, \rho_2)$ is given by (3.26), with the addition of the unknown constant σ^2.

Locally most mean powerful invariant test

In this subsection the null hypothesis

$$H_0: \rho_1 = \rho_2 = 0,$$

is tested against an alternative hypothesis

$$H_a: \rho_1 \geq 0, \rho_2 \geq 0, \text{ (excluding } H_0).$$

Note that the above testing problem is also invariant under the group of transformations (4.2). One can also note that ω, given by (4.3), is a maximal invariant vector, whose joint density is given by (4.4), with the modification that the covariance matrix $\Delta(\rho)$ is replaced by $\Omega(\rho_1, \rho_2)$. Using equation (3) of King and Wu's (1990) theorem and the density of ω, the LMMPI test for the three-stage model is based on the critical region of the form

$$d = \frac{Y'AY}{Y'Y} < c_\alpha,$$

(4.12)

where c is a suitably chosen constant. Note that in (4.12)

$$A = A_1 + A_2,$$

where

$$A_1 = (I_n - D_1),$$

$$A_2 = (I_n - D_2)$$

and the matrices D_1 and D_2 are defined by (3.13) under the restriction that the block and subblock sizes are balanced. More conveniently, d in (4.12) can also be expressed as,

$$
\begin{aligned}
d &= \frac{Y'A_1Y}{Y'Y} + \frac{Y'A_2Y}{Y'Y} \\
&= \frac{Y'(I_n - D_1)Y}{Y'Y} + \frac{Y'(I_n - D_2)Y}{Y'Y} \\
&= 2 - \left[\frac{Y'D_1Y}{Y'Y} + \frac{Y'D_2Y}{Y'Y} \right] \\
&= 2 - \left[\sum_{i=1}^{m} \left(\sum_{j=1}^{s} \sum_{k=1}^{T} y_{ijk} \right)^2 + \sum_{i=1}^{m} \sum_{j=1}^{s} \left(\sum_{k=1}^{T} y_{ijk} \right)^2 \right] / \sum_{i=1}^{m} \sum_{j=1}^{s} \sum_{k=1}^{T} y_{ijk}^2.
\end{aligned}
$$

(4.13)

Note that d in (4.12) and/or in (4.13) is a ratio of quadratic forms in normal variables, therefore, its critical values, c_α, may be obtained by evaluating probabilities of the form,

$$\Pr[d < c_\alpha] = \Pr[Y'(A - c_\alpha I_n)Y < 0]$$

$$= \Pr[Y'\{(2 - c_\alpha)I_n - (D_1 + D_2)\}Y < 0]$$

$$= \Pr\left[\sum_{i=1}^{n} \lambda_i \xi_i^2 < 0\right]$$

$$= \Pr[(2 - c_\alpha)\chi^2_{ms(T-1)} + (2 - c_\alpha - T)\chi^2_{m(s-1)}$$

$$+ (2 - c_\alpha - T(s+1))\chi^2_m < 0], \tag{4.14}$$

where $\xi_i \sim IN(0, 1)$ and λ_i, $i = 1, 2,...,n$, are eigenvalues of the matrix $\{(2 - c_\alpha)I_n - (D_1 + D_2)\}$. The last equality of (4.14) is obtained by noting that the eigenvalues of the latter matrix are $(2 - c_u)$ with multiplicity of $ms(T - 1)$, $(2 - c_\alpha - T)$ with multiplicity of $m(s-1)$ and $\{2 - c_\alpha - T(s+1)\}$ with multiplicity of m, respectively. Also note that (4.14), can be evaluated by standard numerical algorithms mentioned in the previous chapter.

Point optimal invariant test

This subsection reconsiders the testing problem of the previous subsection, i.e. $H_0 : \rho_1 = \rho_2 = 0$ against the simple alternative $H_a : \rho_1 = \rho_{11} > 0, \rho_2 = \rho_{22} > 0$.

Based on the maximal invariant (4.3), King's (1987b) equation (18) gives the POI test of H_0 against H_a. It rejects the null hypothesis for small values of

$$s(\rho_{11}, \rho_{22}) = \frac{Y'\Omega^{-1}(\rho_{11}, \rho_{22})Y}{Y'Y},$$

where $\Omega^{-1}(\rho_{11}, \rho_{22})$ is given by (3.33).

Note that $s(\rho_{11}, \rho_{22})$ is a ratio of quadratic forms in normal variables, therefore its critical values may be obtained by evaluating the probabilities of the form,

$$\Pr[s(\rho_1, \rho_2) < c_\alpha] = \Pr[Y'(\Omega^{-1}(\rho_{11}, \rho_{22}) - c_\alpha I_n)Y < 0]$$

$$= \Pr\left[\sum_{i=1}^{n} \lambda_i \xi_i^2 < 0\right]$$

$$= \Pr\left[\left\{(1-\rho_{11}-\rho_{22})^{-1} - c_\alpha\right\}\chi^2_{ms(T-1)}\right.$$

$$+\left\{(1-\rho_{11}-\rho_{22}(1-T))^{-1} - c_\alpha\right\}\chi^2_{m(s-1)}$$

$$\left.+\left\{(1+\rho_{11}(sT-1)+\rho_{22}(T-1))^{-1} - c_\alpha\right\}\chi^2_m < 0\right],$$

$$(4.15)$$

where $\xi_i \sim IN(0,1)$ and λ_i, $i = 1,\dots,n$, are the eigenvalues of the matrix $(\Omega^{-1}(\rho_{11},\rho_{22}) - c_\alpha I_n)$, namely, $\left\{(1-\rho_{11}-\rho_{22})^{-1} - c_\alpha\right\}$, $\left\{(1-\rho_{11}-\rho_{22}(1-T))^{-1} - c_\alpha\right\}$ and $\left\{(1+\rho_{11}(sT-1)+\rho_{22}(T-1))^{-1} - c_\alpha\right\}$, with multiplicities of ms(T -1), m(s - 1) and m, respectively.

Given a critical value, c_α, for a desired level of significance, α, the power of the critical region $s(\rho_{11}, \rho_{22}) < c_\alpha$ is obtained by evaluating the probabilities of the form,

$$\Pr\left[s(\rho_{11},\rho_{22}) < c_\alpha \middle| Y \sim N(0,\sigma^2\Omega(\rho_1,\rho_2))\right]$$

$$= \Pr\left[Y'\left(\Omega^{-1}(\rho_{11},\rho_{22}) - c_\alpha I_n\right)Y < 0 \middle| Y \sim N(0,\sigma^2\Omega(\rho_1,\rho_2))\right]$$

$$= \Pr\left[z'(\Omega^{1/2})'\left(\Omega^{-1}(\rho_{11},\rho_{22}) - c_\alpha I_n\right)\Omega^{1/2}z < 0\right]$$

$$= \Pr\left[\sum_{i=1}^{n} \lambda_i \xi_i^2 < 0\right],$$

$$(4.16)$$

where $\xi_i \sim IN(0,1), \lambda_i$, $i = 1,\dots,n$, are the eigenvalues of $\left[(\Omega^{1/2})'(\Omega^{-1}(\rho_{11},\rho_{22}) - c_\alpha I_n)\Omega^{1/2}\right]$. The second last equality of (4.16) is obtained by noting that $z = \Omega^{-1/2}Y \sim N(0,I_n)$, where $\Omega = \Omega(\rho_1,\rho_2)$.

It follows from lemma 2 of appendix 3A, that the eigenvalues of

$$\left[\left(\Omega^{1/2}\right)'\left(\Omega^{-1}(\rho_{11},\rho_{22})-c_\alpha I_n\right)\Omega^{1/2}\right]$$

are also the eigenvalues of

$$\left(\Omega^{-1}(\rho_{11},\rho_{22})-c_\alpha I_n\right)\Omega(\rho_1,\rho_2),$$

or equivalently, the eigenvalues of

$$\left\{\Omega^{-1}(\rho_{11},\rho_{22})\Omega(\rho_1,\rho_2)-c_\alpha\Omega(\rho_1,\rho_2)\right\}.$$

The eigenvalues of the latter matrices are obtained by using lemma 4 of appendix 3A. These eigenvalues are

$$a^* = (1-\rho_1-\rho_2)\left[\frac{1}{1-\rho_{11}-\rho_{22}}-c_\alpha\right], \tag{4.17}$$

$$b^* = \{1-\rho_1-\rho_2(1-T)\}\left[\frac{1}{\{1-\rho_{11}-\rho_{22}(1-T)\}}-c_\alpha\right], \tag{4.18}$$

and

$$c^* = \{1-\rho_1(1-sT)-\rho_2(1-T)\}\left[\frac{1}{\{1-\rho_{11}(1-sT)-\rho_{22}(1-T)\}}-c_\alpha\right], \tag{4.19}$$

with multiplicities of $ms(T-1)$, $m(s-1)$ and m, respectively. Thus, given these eigenvalues, the power function (4.16), can be written as

$$\Pr\left[a^*\chi^2_{ms(T-1)}+b^*\chi^2_{m(s-1)}+c^*\chi^2_m<0\right], \tag{4.20}$$

where a*, b* and c* are defined in (4.17), (4.18) and (4.19), respectively.

In the special cases in which power is being evaluated at $(\rho_1,\rho_2)'=(\rho_{11},\rho_{22})'$, (4.17), (4.18) and (4.19) become

$$1-c_\alpha(1-\rho_{11}-\rho_{22}),$$

$$1 - c_\alpha(1 - \rho_{11} - \rho_{22}(1-T)),$$

and

$$1 - c_\alpha\{1 - \rho_{11}(1-sT) - \rho_{22}(1-T)\},$$

respectively. Thus (4.20) can be written as,

$$\Pr\Big[\{1 - c_\alpha(1 - \rho_{11} - \rho_{22})\}\chi^2_{ms(T-1)} + \{1 - c_\alpha(1 - \rho_{11} - \rho_{22}(1-T))\}$$

$$\chi^2_{m(s-1)} + \big[1 - c_\alpha\{1 - \rho_{11}(1-sT) - \rho_{22}(1-T)\}\big]\chi^2_m < 0\Big]$$

$$= \Pr\Big[\chi^2_{ms(T-1)} + d\chi^2_{m(s-1)} + f\chi^2_m < 0\Big], \qquad\qquad (4.21)$$

where

$$d = \frac{\{1 - c_\alpha(1 - \rho_{11} - \rho_{22}(1-T))\}}{\{1 - c_\alpha(1 - \rho_{11} - \rho_{22})\}} \quad \text{and}$$

$$f = \frac{\big[1 - c_\alpha\{1 - \rho_{11}(1-sT) - \rho_{22}(1-T)\}\big]}{\{1 - c_\alpha(1 - \rho_{11} - \rho_{22})\}}.$$

Note that both (4.16) and (4.21) can be evaluated by using standard numerical techniques mentioned in previous sections. Further, using the same arguments as previously stated, it can be easily seen that the $s(\rho_{11},\rho_{22})$ test is a POI for the three-stage SMN distribution model.

POI test for testing subblock effects

In this section the similar testing problem to that of testing (3.41) against (3.42) will be considered under the restriction that Y follows SMN distributions model which is given by (4.11). Note that this testing problem is invariant under the group of transformations of the form (4.2), with maximal invariant vector, ω, given by (4.3). The density of ω, under H_a is given by

$$f_a(\omega, \rho_1, \rho_2) = \frac{1}{2} \Gamma\left(\frac{n}{2}\right) \pi^{-n/2} |\Omega_1|^{-1/2} \left(\omega' \Omega_1^{-1} \omega\right)^{-n/2},$$

while under H_0 it becomes

$$f_0(\omega, \rho_1, 0) = \frac{1}{2} \Gamma\left(\frac{n}{2}\right) \pi^{-n/2} |\Omega_0|^{-1/2} \left(\omega' \Omega_0^{-1} \omega\right)^{n/2}$$

where Ω_0 and Ω_1 are defined earlier in chapter three. Thus, our testing problem (3.41) against (3.42) simplifies to one of testing

$H_0: \omega$ has the density $f_0(\omega, \rho_1, 0)$, $0 \le \rho_1 \le 0.5$,

against

$H_a: \omega$ has the density $f_a(\omega, \rho_1, \rho_2)$, $0 \le \rho_1 \le 0.5$, $0 \le \rho_2 \le 0.5$.

To begin with, the simpler problem of testing

$$H_0': (\rho_1, \rho_2)' = (\rho_{10}, 0)'$$

against the simple alternative hypothesis

$$H_a': (\rho_1, \rho_2)' = (\rho_{11}, \rho_{21})',$$

where the range of $\rho_{10}, \rho_{11}, \rho_{21}$ are the same as defined earlier in chapter three dealing with the similar problem of SSMN model. The Neyman-Pearson lemma yields the critical region of the form

$$s(\rho_{10}, \rho_{11}, \rho_{21}) = \frac{Y' \Omega_{11}^{-1} Y}{Y' \Omega_{10}^{-1} Y} < c_\alpha \tag{4.22}$$

where c_α, in (4.22) is an appropriate critical value. In order to compute c_α, one needs to evaluate the probability of the form,

$$\Pr\left[s(\rho_{10}, \rho_{11}, \rho_{21}) < c_\alpha \middle| Y \sim N(0, \Omega(\rho_{10}, 0)) \right]$$

$$= \Pr\left[Y'\left(\Omega_{11}^{-1} - c_\alpha \Omega_{10}^{-1}\right) Y < 0 \middle| Y \sim N(0, \Omega_0) \right]$$

83

$$= \Pr\left[z'\left(\Omega_0^{1/2}\right)'\left(\Omega_{11}^{-1} - c_\alpha\Omega_{10}^{-1}\right)\Omega_0^{1/2}z < 0 | z \sim N(0,I_n)\right]$$

$$= \Pr\left[\sum_{i=1}^{n}\lambda_i\xi_i^2 < 0\right] \tag{4.23}$$

where $\xi_i \sim IN(0,1), \lambda_i, i = 1,\ldots,n$, are the eigenvalues of

$$\left(\Omega_0^{1/2}\right)'\left(\Omega_{11}^{-1} - c_\alpha\Omega_{10}^{-1}\right)\Omega_0^{1/2},$$

or equivalently, the eigenvalues of

$$\left(\Omega_{11}^{-1} - c_\alpha\Omega_{10}^{-1}\right)\Omega_0 = \Omega_{11}^{-1}\Omega_0 - c_\alpha\Omega_{10}^{-1}\Omega_0,$$

in which $\Omega_{11} = \Omega(\rho_{11},\rho_{22})$, $\Omega_{10} = \Omega(\rho_{10},0)$ and $\Omega_0 = \Omega(\rho_1,0)$.

Using lemmas 2 and 4 of appendix 3A, the eigenvalues of the latter matrix can be obtained as follows,

$$\bar{a} = (1-\rho_1)\left\{\frac{1}{1-\rho_{11}-\rho_{22}} - \frac{c_\alpha}{1-\rho_{10}}\right\}, \tag{4.24}$$

$$\bar{b} = \frac{(1-\rho_1)}{(1-\rho_{11}-\rho_{22})}\left\{\frac{1-\rho_{10}-\rho_{22}}{1-\rho_{11}-\rho_{22}+T\rho_{22}} - \frac{c_\alpha}{1-\rho_{10}}\right\}, \tag{4.25}$$

and

$$\bar{c} = (1-\rho_1+sT\rho_1)\left\{\frac{1}{(1-\rho_{11}-\rho_{22}+T\rho_{22}+sT\rho_{11})}\right.$$

$$\left. -c_\alpha\left(\frac{1}{(1-\rho_{10})(1-\rho_1+sT\rho_1)} - \frac{\rho_{10}}{(1-\rho_{10})(1-\rho_{10}+sT\rho_{10})}\right)\right\}, \tag{4.26}$$

with multiplicities ms(T-1), m(s-1) and m, respectively.

Therefore (4.23) can be written as,

84

$$\Pr\left[\bar{a}\chi^2_{ms(T-1)} + \bar{b}\chi^2_{m(s-1)} + \bar{c}\chi^2_m\right],$$

where $\bar{a}$, $\bar{b}$ and $\bar{c}$ are given by (4.24), (4.25) and (4.26), respectively. Thus a test based on $s\left(\rho_{10},\rho_{11},\rho_{21}\right)$ in (4.22) is most powerful in the neighbourhood of $(\mu_1,\mu_2)'=(\rho_{11},\rho_{21})'$. Therefore it is a POI test of a simple null H_0' against a simple alternative H_a'. For the wider problem of testing (3.41) against (3.42) it may not necessarily be POI, because for this testing problem one needs to solve the probability of the form (3.50), based on the $s\left(\rho_{10},\rho_{11},\rho_{21}\right)$ test for at least one ρ_1 value. If one could succeed to find such a value of $\rho_1 = \rho_{10}$, say, then (4.24), (4.25) and (4.26) become

$$\bar{a}'=\left\{\frac{1-\rho_{10}}{1-\rho_{11}-\rho_{22}} - c_\alpha\right\},$$

$$\bar{b}'=\left\{\frac{(1-\rho_{10})}{(1-\rho_{11}-\rho_{22}+T\rho_{22})} - \frac{c_\alpha}{(1-\rho_{11}-\rho_{22})}\right\}, \text{ and}$$

$$\bar{c}'=\left\{\frac{(1-\rho_{10}+sT\rho_{10})}{(1-\rho_{11}-\rho_{22}+T\rho_{22}+sT\rho_{11})} - c_\alpha\right\}, \text{ respectively.}$$

Thus the critical values can be obtained by solving the left hand side of

$$\Pr\left[\bar{a}'\chi^2_{ms(T-1)} + \bar{b}'\chi^2_{m(s-1)} + \bar{c}'\chi^2_m < 0\right] = \alpha,$$

where α is the desired significance level. The exact critical values and the power calculations for the similar testing problem would be considered in chapter six, while dealing with the 3SLR model in which disturbances follow these distributions.

Finally, the LMMPI and the POI tests can easily be generalized to a multi-stage block design for testing block, sub-block and/or sub-sub-block equicorrelation coefficients, whether these correlation coefficients are equal to zero or they have some positive values, within blocks, sub-blocks and/or sub-sub-blocks, respectively. Discussion on some of these more complicated models will be considered in chapter six of this book.

Concluding remarks

In this chapter the POI and LMMPI tests for the case of two- and three-stage SMN distribution model was discussed. These tests have been developed in the presence of the nuisance parameter, σ^2, which has been eliminated by using an invariance technique. The most attractive feature of the POI and LMMPI tests for the case of the two-stage SMN distribution model is that their power and the critical values are easily obtainable from the standard F distribution. An additional feature of the POI test is that it is UMPI.

LMMPI and POI tests for the case of three-stage SMN model for testing whether $\rho_1 = \rho_2 = 0$ against the alternative that they have some positive value were derived. The POI test for testing ρ_2 in the presence of $\rho_1 > 0$ was also developed. A brief summary of the tests developed in the previous chapter and in this chapter are given in tables 4.1 and 4.2 of appendix 4A. There are some other topics worth exploring when using a model which follows a SSMN or SMN distribution. One is the question of testing block and subblock effects in the context of the multi-stage linear regression model, when disturbances follow these distributions. These problems will be discussed in the subsequent chapters of this book.

Appendix 4A: Summary of the tests of the SSMN and SMN models

Table 4.1

Some optimal tests for two-stage SSMN and SMN distributions

Name of tests	Null hypothesis	Alternative hypothesis	Form of covariance matrix	Test statistics
SSMN				
1 LB	$\rho=0$	$\rho>0$	$\Delta(\rho_0)$	$s = Y'(I_n - D)Y < c_\alpha$
2 BO	$\rho=0$	$\rho>0$	$\Delta(\rho_1)$	$r(\rho_1) = Y'(\Delta^{-1}(\rho_1) - I_n)Y < c_\alpha$
3 LMMP	$\rho_i=0$ $i=1,...,m$	$\rho_i \geq 0$ $i=1,...,m$	$\Delta(\rho_i)$	$d = Y'(I_n - D)Y < c_2$
SMN				
4 POI	$\rho=0$	$\rho>0$	$\sigma^2\Delta(\rho_1)$	$s(\rho_1) = \dfrac{Y'\Delta^{-1}(\rho_1)Y}{Y'Y} < c_\alpha$
6 LMMPI	$\rho_i=0$ $i=1,...,m$	$\rho_i \geq 0$ $i=1,...,m$	$\sigma^2\Delta(\rho_i)$	$s = \dfrac{Y'(I_n - D)Y}{Y'Y} < c_2$

Table 4.2

Some optimal tests for three-stage SSMN and SMN distributions

Name of tests	Null hypothesis	Alternative hypothesis	Form of covariance matrix	Test statistics
SSMN				
1 LMMP	$\rho_1=\rho_2=0$	$\rho_1 \ge 0, \rho_2 \ge 0$	$\Omega(\rho_1,\rho_2)$	$d^* = Y'\{2I_n - (D_1+D_2)\}Y < c_3$
2 PO	$\rho_1=\rho_2=0$	$\rho_1=\rho_{11}>0$ $\rho_2=\rho_{22}>0$	$\Omega(\rho_{11},\rho_{22})$	$r(\rho_{11},\rho_{22}) = Y'\left(\Omega^{-1}(\rho_{11},\rho_{22}) - I_n\right)Y < c_\alpha$
3 PO	$\rho_2=0,\rho_1>0$	$\rho_1=\rho_{11}>0$	under $H_0: \Omega_{10}=\Omega(\rho_{10}.0)$ under $H_a: \Omega_{11}=\Omega(\rho_{11},\rho_{21})$	
(for sub-block effects)		$\rho_2=\rho_{21}>0$		$r(\rho_{10},\rho_{11},\rho_{21}) = Y'(\Omega_{11}^{-1} - \Omega_{10}^{-1})Y < c_\alpha$
SMN				
4 LMMPI	$\rho_1=\rho_2=0$	$\rho_1 \ge 0, \rho_2 \ge 0$	$\sigma^2\Omega(\rho_1,\rho_2)$	$d = 2 - \dfrac{Y'(D_1+D_2)Y}{Y'Y} < c_\alpha$
5 POI	$\rho_1=\rho_2=0$	$\rho_1=\rho_{11}>0$ $\rho_2=\rho_{22}>0$	$\sigma^2\Omega(\rho_{11},\rho_{22})$	$s(\rho_{11},\rho_{22}) = \dfrac{Y'\Omega(\rho_{11},\rho_{22})Y}{Y'Y} < c_\alpha$
6 POI	$\rho_2=0,\rho_1>0$	$\rho_1=\rho_{11}>0$	$\sigma^2\Omega_{10}$	
(for sub-block effects)		$\rho_2=\rho_{21}>0$	$\sigma^2\Omega_{11}$	$s(\rho_{10},\rho_{11},\rho_{21}) = \dfrac{Y'\Omega_{11}Y}{Y'\Omega_{10}Y} < c_\alpha$

5 Testing for block effects in two-stage linear regression models

Introduction

It is common practice for survey data to be collected in blocks or clusters and the standard linear regression model to be fitted to such data, with the disturbances assumed to be mutually independent. This practice entails a number of potential pitfalls, which have been discussed in the first two chapters. In this chapter, hypothesis testing problems associated with the linear regression model arising from two-stage block survey data is considered. To recall Deaton and Irish (1983)where he, considered a regression model based on a two-stage blocked (or clustered) sampling design and suggested the use of a one-sided Lagrange multiplier (LM1) test, whereas King and Evans (1986) showed that the test is LBI for testing for a non-zero equicorrelation coefficient, ρ, within the blocks.

The main purpose of this chapter is to construct a POI test and to investigate whether the power of this test is better than that of the LBI test, for small and moderate sample sizes. It is found that for testing $H_0:\rho=0$, against $H_a:\rho>0$, both the LBI and POI tests are approximately UMPI, for some selected block sizes and for these cases the critical values of this test can be approximated by critical values from the standard F distribution. The estimated sizes based on these critical values have the tendency to become closer to the nominal size of 0.05, when the sample size and the value of ρ_1 increase.

Finally, this chapter derives a locally mean most powerful invariant (LMMPI) test for testing the hypothesis, $\rho = 0$, against the alternative that ρ varies from block to block and finds that the LMMPI test is equivalent to that of the LM1 test.

The content of this chapter is divided into the following sections. In the subsequent sections, the two-stage linear regression model is introduced and a procedure for constructing a POI test for testing $\rho = 0$, against a positive value of ρ is outlined. The application of King and Wu's (1990) LMMPI test to more complicated models is also demonstrated, in which it is assumed that the equicorrelation coefficients vary from block to block and show that it is equivalent to the LM1 test. In the next section, the critical values of the POI test is approximated using the standard F distribution for selected block sizes and it presents the five versions of the POI tests which optimize power at $\rho = 0.1, 0.2, 0.3, 0.4$ and 0.5, respectively. It is found that the $s(0.3)$ test, which optimizes power at $\rho = 0.3$, is the best overall test. It also reports an empirical power comparison of the $s(0.3)$, LM1, LM2, Durbin-Watson (DW), and King's (1981) modified DW tests with that of the power envelope (PE). The final section contains some concluding remarks.

Two-stage linear regression models and the tests

Theory

Following Scott and Holt (1982), King and Evans (1986) and the 2SLR model (2.1) of chapter two, it is assumed that n observations are available from a two-stage sample with m blocks or clusters. Let $m(i)$ be the number of observations from the ith block so that $n = \sum_{i=1}^{m} m(i)$. It is assumed that the data are ordered so that the observations within a block are adjacent. Therefore, the model in detail is expressed as

$$y_{ij} = \sum_{k=1}^{p} \beta_k x_{ijk} + u_{ij}$$

(5.1)

where

$$u_{ij} = v_i + v_{ij}, \quad i = 1,...,m, \quad j = 1,2,...,m(i),$$

in which i is the block number, j is the observation number in the given block, β_k are unknown constant coefficients and x_{ijk} for $k = 1,2,...,p$ are observations on p independent variables, the first of which is a constant. The error terms u_{ij} have two components: v_i, the ith random shock, specific to the ith block and the remaining part, v_{ij}, the individual shocks. Hence it is assumed,

1 v_i and v_{ij} are normal and mutually independent,

2 $E(v_i) = E(v_{ij}) = 0$, for $i = 1,...,m$, $j = 1,2,...,m(i)$,

3 $E(v_i v_{i'}) = \begin{cases} \sigma_1^2, & i=i' \\ 0, & \text{otherwise,} \end{cases}$

4 $E(v_{ij} v_{i'j'}) = \begin{cases} \sigma_2^2, & i=i', \ j=j' \\ 0, & \text{otherwise.} \end{cases}$

The above assumptions (1 to 4) imply that:

$E(u_{ij}) = 0$, for all i and j;

$$E(u_{ij}\, u_{st}) = \sigma^2 \delta_{is}\left\{\rho + (1-\rho)\delta_{jt}\right\}, \tag{5.2}$$

where $\rho = \dfrac{\sigma_1^2}{\sigma^2}$, such that $0 \le \rho \le 1$, $\sigma^2 = \sigma_1^2 + \sigma_2^2$, $\delta_{ii} = 1$ and $\delta_{ij} = 0$, for $i \ne j$.

In matrix form, (5.1) can be written as

$$y(i) = X(i)\beta + u(i); \tag{5.3}$$

for $i = 1,...,m$, where $y(i)$ is an $m(i) \times 1$ vector of y_{ij} values, $X(i)$ is an $m(i) \times p$ matrix whose (j,k)th element is x_{ijk} and $u(i)$ is an $m(i) \times 1$ vector of u_{ij} values. If y and u denote the $n \times 1$ vectors of stacked $y(i)$ and $u(i)$ vectors respectively, and X is the $n \times p$ stacked matrix of $X(i)$ matrices, then (5.3) can be written as

$$y = X\beta + u \tag{5.4}$$

such that

$$u \sim N\left(0, \sigma^2 \Delta(\rho)\right),$$

where $\Delta(\rho) = \overset{m}{\underset{i=1}{\oplus}} \Delta_i(\rho)$ is block diagonal, with submatrices

$$\Delta_i(\rho) = (1-\rho)I_{m(i)} + \rho E_{m(i)} \qquad (5.5)$$

in which $I_{m(i)}$ is an $m(i) \times m(i)$ identity matrix and $E_{m(i)}$ is $m(i) \times m(i)$ matrix with all elements equal to one. Although we can proceed with unequal block sizes in this model, for the sake of simplicity in exposition, we will consider the case of balanced data, (i.e., equal block sizes) and hence we assume that $m(i)=k$ and $n=km$ in the subsequent sections. With this notation, one can note that (5.5) is equivalent to (3.1) and hence $\Delta(\rho)$, in this chapter is the same as in (3.4) of chapter three.

Point optimal invariant test

In this subsection we construct the MPI test statistic for testing $H_0: \rho = 0$ against the specific alternative $H_a': \rho = \rho_1 > 0$, in the context of model (5.1), and under the assumptions (1 to 4). This statistic forms the basis of a point optimal test for the wider testing problem.

Observe that the problem of testing $H_0: \rho=0$, against $H_a: \rho>0$ is invariant under the group of transformations,

$$y \rightarrow \gamma_0 y + X\gamma, \qquad (5.6)$$

where γ_0 is a positive scalar and γ is $p \times 1$ vector. Let $M_x = I_n - X(X'X)^{-1}X'$, $\hat{u} = M_x y = M_x u$ be the OLS residual vector from (5.4) and P be an $(n-p) \times n$ matrix such that $PP' = I_{(n-p)}$ and $P'P = M_x$. The vector

$$\vartheta = \frac{P\hat{u}}{(\hat{u}'P'P\hat{u})^{1/2}}$$

is a maximal invariant under the group of transformations given by (5.6). Based on ϑ, we choose $\rho=\rho_1$, a known point at which we want to have optimal power in the alternative parameter space. By using King's (1987b), equation (18), we can obtain the POI test, which is to reject the null hypothesis for small values of

$$s(\rho_1) = \frac{\hat{u}'A\hat{u}}{\hat{u}'\hat{u}},$$

92

$$= \frac{u'Au}{u'M_x u},$$ (5.7)

where

$$A = \Delta_1^{-1} - \Delta_1^{-1}X\left(X'\Delta_1^{-1}X\right)^{-1}X'\Delta_1^{-1},$$

$$\Delta_1 = \Delta(\rho_1).$$

The last equality of (5.7) is obtained by observing that $\hat{u}'A\hat{u} = u'M_x AM_x u = u'Au$. As King (1987a, p. 23) notes, $s(\rho_1)$ in (5.7) can also be written as

$$\frac{\sum_{i=1}^{n-p} \lambda_i \xi_i^2}{\sum_{i=1}^{n-p} \xi_i^2}$$ (5.8)

where $\lambda_1, \ldots, \lambda_{n-p}$ are the reciprocals of the non-zero eigenvalues of $\Delta_1 M_x$ or, equivalently, the non-zero eigenvalues of A. Thus, for any given value of ρ_1, at the desired level of significance, α, the critical value, c_α, can be found by solving

$$Pr\left[\sum_{i=1}^{n-p} (\lambda_i - c_\alpha)\xi_i^2 < 0\right] = \alpha.$$ (5.9)

This can be done by using either Koerts and Abrahamse's (1969) FQUAD subroutine or Davies' (1980) versions of Imhof's (1961) algorithm or Farebrother's (1980) version of Pan's (1968) procedure to evaluate the left-hand-side of (5.9). More recently, Shively, Ansley and Kohn (1990) and Palm and Sneek (1984) have suggested alternative procedures to compute (5.9) which do not require the computation of the eigenvalues.

Once the critical values have been found, one may wish to calculate the power of the $s(\rho_1)$ test. To do so, one need to decompose the error covariance matrix $\Delta(\rho)$, or its diagonal component matrices $\Delta_i(\rho)$, in (5.5). This can be done by noting that the Cholesky decomposition of $\Delta(\rho)$ is

93

$$\Delta(\rho) = T(\rho)T'(\rho)$$

where $T(\rho)$ can be obtained through S, via the recursive scheme given in (3.9). Observe that the power of the critical region $s(\rho_1) < c_\alpha$, in (5.7) can be found by evaluating the probabilities of the form,

$$\Pr(s < c_\alpha) = \Pr[u'(A - c_\alpha M_x)u < 0],$$

where

$$u \sim N(0, \sigma^2 \Delta(\rho)).$$

Let $\xi = T(\rho)^{-1}u \sim N(0, \sigma^2 I_n)$, so that $u = T(\rho)\xi$. Substituting for u in (5.10), one can obtain the power function of the form

$$\Pr[\xi'T'(\rho)(A - c_\alpha M_x)T(\rho)\xi < 0]$$

$$= \Pr[\xi'F\xi < 0]$$

$$= \Pr\left[\sum_{i=1}^{n}\lambda_i\eta_i^2 < 0\right],$$

where the λ_i's are the eigenvalues of $F = T'(\rho)(A - c_\alpha M_x)T(\rho)$ and the η_i^2's are independent chi-squared random variables each with one degree of freedom. It is worth noting that under H_a the matrix A in (5.7), can be expressed as $A = (T^{-1}(\rho_1))'M^*T^{-1}(\rho_1)$, where $M^* = I - X^*(X^{*'}X^*)^{-1}X^{*'}$ and $X^* = T^{-1}(\rho_1)X$. The block diagonal lower triangular matrix $T^{-1}(\rho)$ is of the form [1]

$$T^{-1}(\rho) = \begin{bmatrix} Q_k & 0 & \cdots & & 0 \\ 0 & Q_k & \cdots & & \cdot \\ \cdot & & & & \\ \cdot & & & & \cdot \\ \cdot & & & \cdot & \\ 0 & \cdots & & 0 & Q_k \end{bmatrix}$$

94

where Q_k is a kxk lower triangular matrix whose elements are obtained by the formula given below:

$$q_{ii} = \sqrt{\frac{[1+(i-2)\rho]}{(1-\rho)[1+(i-1)\rho]}}$$

and

$$q_{ij} = \frac{-\rho}{\sqrt{(1-\rho)[1+(i-1)\rho][1+(i-2)\rho]}}$$

for $i = 1, 2,...,k$; $j = 2, 3,...,k$ and $j > i$. The last equality of (5.12) can be calculated by using any of the methods outlined above for calculating (5.9).

Locally most mean powerful invariant test

In previous subsections a POI test has been constructed for the equicorrelation coefficient, ρ, which was based on the error covariance matrix (5.5) of model (5.4), and involves only one parameter, ρ, in all the blocks. In this subsection a more general situation is considered which is based on the discussion in chapter three of the section entitled, a more complicated model and the LMMP test, where it is assumed that the equicorrelation coefficient varies from block to block or from cluster to cluster. Therefore, in such cases the subblocks of the error covariance matrix (5.5) can be expressed in the form of (3.5), i.e.

$$\Delta_i(\rho_i) = (1-\rho_i)I_k + \rho_i E_k,$$

for $i = 1,2,...,m$.

Our problem of interest is to test

$$\overline{H}_0 : \rho_1 = \rho_2 = ... = \rho_m = 0 \tag{5.12}$$

against

$$\overline{H}_a : \rho_1 \geq 0, \rho_2 \geq 0,...,\rho_m \geq 0, \quad \text{(excluding } H_0\text{)}. \tag{5.13}$$

Note that this testing problem is invariant under the group of transformations given by (5.6).

95

King and Wu (1990) showed that a LMMPI test for $\overline{H}_0$ against $\overline{H}_a$ is to reject $\overline{H}_0$ for small values of

$$d' = \frac{\hat{u}'A * \hat{u}}{\hat{u}'\hat{u}}, \qquad (5.14)$$

where

$$A* = \sum_{j=1}^{m} A_j,$$

$$A_i = -\frac{\partial\Delta(\rho)}{\partial\rho_i}\bigg|_{\rho=0},$$

for i=1,...,m, and $\hat{u}$ is a vector of OLS residual from (5.4).

For our problem of testing (5.12) against (5.13), the LMMPI test is determined by (5.14) with $A = (I_n - D)$, which is given by (3.16), where $D = I_m \otimes E_k$. Substituting A in (5.14), we can express d´ as,

$$d' = \frac{\hat{u}'(I_n - D)\hat{u}}{\hat{u}'\hat{u}}$$

$$= 1 - \frac{\hat{u}'D\hat{u}}{\hat{u}'\hat{u}}$$

$$= 1 - \frac{\sum_{i=1}^{m}\left(\sum_{j=1}^{k}\hat{u}_{ij}\right)^2}{\sum_{i=1}^{m}\sum_{j=1}^{k}\hat{u}_{ij}^2}. \qquad (5.15)$$

It is worth noting that d´ is the LBI test of $\overline{H}_0$ in the direction of $\rho_1 = \rho_2 = ... = \rho_m > 0$, (see King and Wu (1990) for a property of the LMMPI test) and also note that d´ test is equivalent to the LM1 test of H_0 against H_a introduced by Deaton and Irish (1983). Observe that the form of equation (5.15) is a ratio of quadratic forms in normal variables, and therefore its critical values may be determined by the standard numerical techniques mentioned earlier. Under H_0, the LM1 test statistic has a standard

96

normal asymptotic distribution while the LM2 test statistic is asymptotically distributed as $\chi^2_{(1)}$. In the next section we will compare the powers of these tests with that of the PE and the existing tests.

An empirical comparison of sizes and powers

The aim of this section is to explore the small-sample power properties of the $s(\rho_1)$ test of H_0 against H_a when the intra-block or equicorrelation coefficient is present in the two-stage linear regression model. The power functions of the $s(\rho_1)$ tests were evaluated in the context of model (5.4) and (5.5) and were compared with that of the DW (d), King's (1981) modified DW (d*), the one sided and two-sided Lagrange multiplier (LM1 and LM2) tests, (as calculated and reported by King and Evans, 1986). Exact five percent critical values of the $s(\rho_1)$ test and the associated values of the power at $\rho = 0.1, 0.2, 0.3, 0.4$ and 0.5 were computed by using a modified version of the FQUAD subroutine of Koerts and Abrahamse (1969), with maximum integration and truncation errors of 10^{-6}.

The X matrices used by King and Evans are based on cross-sectional Australian census data, classified into 64 demographic groups according to 8 categories of sex/marital status and 8 categories of age (see Williams and Sams (1981) for detailed discussion of the data). For these categories the regressors are population, number of households, and headship ratios of households. The first three data sets consisted of a constant dummy and all combinations of two of these three regressors for 1961 and, when extra observations were necessary, 1976. The other two design matrices were composed of a constant and all three regressors for 1961 and 1976, and for 1966 and 1971, respectively. In each case, 32, 64 and 96 observations were used, each with equal block sizes of k=2,4,8 and $\frac{n}{2}$ (i.e., the block size equivalent to half of the sample size). It was also assumed that the data used in this analysis were ordered so that observations from the same block were together. The values of the variables used in this study are presented in tables 5.1, 5.2 and 5.3 of appendix 5A for ready reference and further research.

For a block size of $\frac{n}{2}$ our results appear to show that the critical region of the $s(\rho_1)$ test is approximately invariant to the choice of ρ_1 value and hence it is approximately UMPI. To investigate this possibility one can re-examine the $s(\rho_1)$ test.

97

Using (5.8) the critical region of the $s(\rho_1)$ test can also be written as,

$$\frac{\sum\limits_{i=1}^{n-p} \lambda_i \xi_i^2}{\sum\limits_{i=1}^{n-p} \xi_i^2} < c_\alpha, \tag{5.16}$$

where in this case the λ_i's are the eigenvalues of the matrix

$$A = \Delta_1^{-1} - \Delta_1^{-1} X \left(X' \Delta_1^{-1} X \right)^{-1} X' \Delta_1^{-1},$$

or equivalently of

$$\left[I_n - \left(\Delta_1^{-\frac{1}{2}} \right)' X \left(X' \Delta_1^{-1} X \right)^{-1} X' \left(\Delta_1^{-\frac{1}{2}} \right) \right] \Delta_1^{-1}.$$

By using King's (1987a, p.23) Corollary and a knowledge of the eigenvalues of Δ_1^{-1} from lemma 1 of appendix 3A, we have the following relationships,

$$\begin{cases} \lambda_i = \dfrac{1}{\{1 + (k-1)\rho_1\}}, & \text{for } i = 1, \ldots, m - p \\[3mm] \dfrac{1}{\{1 + (k-1)\rho_1\}} \leq \lambda_i \leq \dfrac{1}{1 - \rho_1}, & \text{for } i = m - p + 1, \ldots, m \\[3mm] \lambda_i = \dfrac{1}{1 - \rho_1}, & \text{for } i = m + 1, \ldots, n - p. \end{cases} \tag{5.17}$$

If m>p, then (5.16) can be expressed as,

$$\frac{\left\{\dfrac{1}{1+(k-1)\rho_1}\right\}\displaystyle\sum_{i=1}^{m-p}\xi_i^2 + \displaystyle\sum_{i=m-p+1}^{m}\lambda_i\xi_i^2 + \dfrac{1}{1-\rho_1}\displaystyle\sum_{i=m+1}^{n-p}\xi_i^2}{\displaystyle\sum_{i=1}^{n-p}\xi_i^2} < c_\alpha. \qquad (5.18)$$

In this case note that when m is small and n is large, the last term of the numerator of (5.18) dominates and hence, $(1-\rho_1)s(\rho_1)$ approximately does not depend on ρ_1. This implies that the $s(\rho_1)$ test for $m > p$ is approximately UMPI.

If $m \leq p$, then (5.18) becomes

$$\frac{\displaystyle\sum_{i=1}^{n-p}\lambda_i\xi_i^2 + \dfrac{1}{1-\rho_1}\displaystyle\sum_{i-m+1}^{n-p}\xi_i^2}{\displaystyle\sum_{i=1}^{n-p}\xi_i^2} < c_\alpha, \qquad (5.19)$$

where the eigenvalues of the numerator of the first term in (5.19) are given by (5.17), i.e. $\dfrac{1}{\left[1+(k-1)\rho_1\right]} \leq \lambda_i \leq \dfrac{1}{1-\rho_1}$, for $i=1,...,m$.

Computationally exact values of these eigenvalues for $m = 2$ were computed by using the numerical algorithm TRED2 from Martin et $al.$ (1968). For each of King and Evans' data sets, it was found that all of the eigenvalues are, $\dfrac{1}{1-\rho_1}$ except for one, which is very small and is approximately equal to $\dfrac{1}{\left[1+(k-1)\rho_1\right]}$. If we ignore the smallest eigenvalue, then the critical regions (5.19) become

$$s^* = \frac{\displaystyle\sum_{i=2}^{n-p}\xi_i^2}{\displaystyle\sum_{i=1}^{n-p}\xi_i^2} < c_\alpha(1-\rho_1). \qquad (5.20)$$

From (5.20) it is clear that the test is approximately independent of the values of ρ_1 and hence it is almost invariant to the choice of ρ_1. Therefore the test based on $s(\rho_1)$ for $m \leq p$ is an approximate UMPI. This also includes the case of block size, $k = \dfrac{n}{2}$ or equivalently, $m = 2$, which makes this approximation stronger. This can also be verified by looking at the last blocks of the tables 5.7 to 5.15 in appendix 5B, where the power of the $s(\rho_1)$ test for the block size $\dfrac{n}{2}$ is unchanged up to the four decimal places. Hence we have the following proposition for the $s(\rho_1)$ test.

Proposition 5.2

The $s(\rho_1)$ test is approximately UMPI for all values of p and m.

It is worth noting that in (5.20), s* can be expressed in terms of the standard F distribution as follows

$$\Pr\left[s^* < c_\alpha(1-\rho_1)\right] = \Pr\left[\frac{1}{s^*} > \frac{1}{c_\alpha^*(1-\rho_1)}\right]$$

$$= \Pr\left[\frac{\xi_1^2}{\sum\limits_{i=2}^{n-p}\xi_i^2} < \left(\frac{1}{c_\alpha^*(1-\rho_1)} - 1\right)\right]$$

$$= \Pr\left[\frac{\chi_{(1)}^2}{\chi_{(n-p-1)}^2/(n-p-1)} > \left(\frac{1}{c_\alpha^*(1-\rho_1)} - 1\right)(n-p-1)\right]$$

$$= \Pr\left[F(\vartheta_1,\vartheta_2) > F_\alpha\right],$$

where

$$F_\alpha = \left(\frac{1}{c_\alpha^*(1-\rho_1)} - 1\right)(n-p-1),$$

$$\vartheta_1 = 1, \quad \vartheta_2 = n-p-1$$

and $F(\vartheta_1, \vartheta_2)$ denotes the F distribution with ϑ_1 and ϑ_2 degrees of freedom. If F_α is the $100(1-\alpha)$ percentile of the F distribution (obtainable from standard Tables), then the approximate critical value, c_α^*, can be obtained by the following equation

$$c_\alpha^* = \frac{1}{(1-\rho_1)\left(\dfrac{F_\alpha}{(n-p-1)} + 1\right)}.$$

Approximation of c_a^* and sizes using F distribution

A comparison of the approximate critical values, c_α^*, and the exact critical values, c_α, for different values of n, $k = \dfrac{n}{2}$ (and/or m=2) at five percent significant level is given in tables 5.4, 5.5 and 5.6, of appendix 5B, using cross-sectional Australian census data of appendix 5A for the years (1961 and 1976), (1966 and 1971) and artificially generated log-normal variable data, respectively. The last data set has been used as a proxy for cross-sectional data in many empirical studies. Examples include, Goldfeld and Quandt (1965), Harvey and Phillips (1974), Harrison and McCabe (1979), Prakash (1979), Harrison (1980) and Buse (1980), among others. The results reported in the last column of the tables 5.4 to 5.6 are the estimated sizes of the POI test, which is based on approximate critical values obtained using the standard F distribution. The exact values of the F distribution with ϑ_1, and ϑ_2 degrees of freedom are obtained by using the CDFFC subroutine of GAUSS system version 2.0. It is worth noting that the results of the exact critical values, c_α and the approximate critical values, c_α^* are almost similar in tables 5.4 and 5.5. This is because the same households have been used in the samples with different time periods. With the change in data set it can be noted that these results are a bit different in table 5.6 because here a data set of a different nature has been considered.

The results reveal that in all cases the estimated sizes of the $s^*(\rho_1)$ test are less than the nominal level and greater than half of the nominal level, 0.05, even in small samples, except for n=32, ρ_1=0.1,0.2 and for n=64, ρ_1=0.1. It can also be noticed in tables 5.4 to 5.6 that there is a tendency for the estimated sizes of the $s^*(\rho_1)$ test to become closer to the nominal sizes when the sample size and the value of ρ_1 increase. This is one of the desirable

101

properties of the POI test it always has sizes over the null parameter space closer to the nominal significance level 0.05. This test has this property for large and moderate sample sizes. An advantage of this is that applied econometricians can use the standard F distribution to calculate approximate critical values of the requisite POI test.

In the rest of this section further investigation has been divided into two parts. The first part reports the power calculations for the five $s(\rho_1)$ tests with $\rho_1 = 0.1, 0.2, 0.3, 0.4$ and 0.5, respectively. After comparing the power performance of these tests, the test whose power is the closest to the power envelope (PE) has been chosen. The next subsection compares the power performance of a selected $s(\rho_1)$ test with that of the LM1, LM2, and d* tests from the King and Evans study and the PE.

Power comparisons of $s(\rho_1)$ tests

Selected results of the calculations of the power of the first part are given in tables 5.7 to 5.15 of appendix 5B, with the five percent critical values, c_α, in column one. In general, under H_a, the power function of all tests increases with the increase in sample size, block size and the value of ρ.

As noted above, for a block of size two, all of the $s(\rho_1)$ tests are approximately UMPI tests because they are close to the PE. For a block size of $\frac{n}{2}$, they are almost UMPI tests because their power is very close to the PE, (as far as these design matrices are concerned) since the critical regions for this particular block size are almost invariant to the choice of ρ_1 values. For example, consider all of the $s(\rho_1)$ tests, for $\rho_1 = 0.1$ to 0.5, for $n = 32$ and block of size two. It is found that the power curves of all tests are approximately identical to the PE for all values of $\rho \leq 0.5$. The maximum power difference of these tests lies in the interval of 0.0001 to 0.0007. This difference is negligible for lower values of ρ and/or large sample sizes.

For the comparison of higher blocks of sizes 4, 8 and 16, the s(.1), s(.2), s(.3), s(.4) and s(.5) tests are MPI in the neighbourhood of $\rho = 0.1, 0.2, 0.3, 0.4$ and 0.5, respectively. The choice between the s(.1) to s(.5) tests depends upon the researcher's requirement. If a researcher is interested in detecting larger values of ρ in smaller sample sizes then perhaps the s(.3), s(.4) or s(.5) tests could be favoured. If one is interested in detecting smaller values of ρ, then the s(.1) or s(.2) tests could be more suitable. For routine use, it is prefered to utilise the s(.3) test, since under H_a the value of $\rho_1 = 0.3$ seems to be more practicable and the calculations of its powers and sizes is the closest to the PE and the nominal size of 0.05, respectively, as compared to other $s(\rho_1)$ tests under consideration. With the increase in the sample size this maximum difference between the PE and the power of s(.3) tests reduces and with the higher block sizes it increases slightly.

Another procedure for comparing the powers of such tests is given by Cox and Hinkley (1974, p. 102). Their suggestion is to choose a test which maximizes some weighted average of the powers. The problem with this procedure is how to determine the appropriate weighted average, as the selected weights may vary from one person to another. To overcome this problem, Brooks' (1992) suggested taking a simple average of the powers. Following Brooks, we have considered a simple average of the powers over a set of ρ values (i.e., $\rho = 0.1, 0.2, 0.3, 0.4, 0.5$) and then compared these between tests.

The average power of these tests is given in the last column of tables 5.7, 5.10 and 5.11 for the design matrices of (1961 and 1976) whereas the results of tables 5.8, 5.11 and 5.14 are obtained from the data of (1966 and 1971). Tables 5.9, 5.12, and 5.15 shows the results for the artificially generated data. This procedure also favours our selection of the s(.3) test for moderate sample sizes. The exceptional cases are for n=32, k=2, 4, n=64, k=16, and n=96, k=4, 8, where the average of the power of the s(.4) and s(.2) tests, respectively, leads to that of the s(.3) test by 0.0001. This difference in average power is very small, therefore it is relatively unimportant for practical purposes. The PE calculated for different values of n and ρ are given in tables 5.16 to 5.21 for the first two design matrices of Australian cross-sectional data. The power patterns of the third data matrix are quite similar to the previous ones. Therefore their results are omitted here. Hence in the next part of this investigation, the power of the s(.3) test will be compared with that of the LM1, LM2, d and d* tests. Here some of the power calculations are reproduced from King and Evans' (1986) table, so that the second part of our investigation should be self-contained.

Power comparisons of s(.3), LM1, LM2, d* and d tests with PE

The numerical results for the powers of the s(.3), LM1, LM2, d* and d tests are tabulated in tables 5.16 to 5.21, for different sample sizes, i.e. n = 32, 64 and 96, respectively. These results demonstrate that the powers of all tests increase as n, k and ρ increase, while other variables remain the same. As expected, the s(.3) test is always the most powerful of the four tests for almost all values of ρ, followed by the LM1, LM2, d* and d tests, respectively. The power difference between the s(.3) and LM1 tests for smaller block sizes is very small. For example, consider the block size of 2 in tables 5.16 to 5.21. It is found that the power of both the tests are approximately identical; with the increase in the sample sizes both the power curves merge. Hence it is confirmed that the LM1 test is a good approximation to the s(.3) test for this special case. Therefore, the LM1 test is also an approximately UMPI test as is the case with the s(.3) test, at least for these design matrices.

For the higher block sizes, the power of the s(.3) test is superior to that of the LM1 test for all values of $\rho \geq 0.2$. This can be verified by considering n = 32, k = 4, where the maximum power difference between the LM1 and s(.3) tests varies from 0.001 to 0.017 and for k = 8, it varies from 0.006 to 0.028. As the sample size, block size and the value of ρ increase, this maximum difference between the powers becomes smaller and smaller and also the average powers of the s(.3) and LM1 tests gets closer and closer. For example, consider the case of n = 96, k = 8. For $0.2 \leq \rho \leq 0.5$, the power difference between the two tests varies from 0.003 to 0.001 and the average power difference is 0.002. Also note that for this particular case the power of the s(.3) test is exactly identical to that of the PE. On the basis of these facts it may be concluded that the POI test is marginally better than the LM1 test at least for small and moderate sample sizes.

As King and Evans noted, the Durbin-Watson (d) test is the least powerful among all the tests under comparison, though it still has reasonable power in the larger samples and the smaller block sizes. The advantages of using the d test over all other tests are, firstly, it has reasonable small sample power properties when the block structure is unavailable, secondly, its computations can easily be done by standard regression packages, with the availability of existing bounds in the econometrics literature. Further details of the comparisons of these existing tests can be found in King and Evans (1986) article.

Concluding remarks

This chapter considered the two-stage linear regression model with equicorrelated disturbances and derived POI tests for the hypothesis that the equicorrelation coefficient, ρ, has any given value. The power of the POI test was compared with that of the existing tests and the PE. The results as detailed in Appendix 5B suggest that the POI test is marginally better than the LM1 test for small and moderate sample sizes, at least for the data sets used in this experiment. It was found that both the tests are approximately UMPI for some selected block sizes. It also showed that for all values of m and p, the POI test is approximately UMPI. Further, the critical values of the POI test can also be obtainable from the standard F distribution for some selected block sizes. The approximate critical values using the F distribution was calculated. It was found that 87% of the cases the estimated sizes based on these critical values are greater than half of the nominal sizes. A LMMPI test for testing the hypothesis that ρ has different values for each block has been derived and it was found that the LMMPI test is equivalent to the LM1 test. In the next chapter a testing procedure for the multiple block effects will be considered and the question of how close the LMMPI test is to that of the PE in the case of 3SLR model will be discussed.

Note

1   Note that $T^{-1}(\rho)$ obtained by Cholesky decomposition of the covariance matrix, $\Delta^{-1}(\rho) = \left(T^{-1}(\rho)\right)' T^{-1}(\rho)$ is not unique.

Appendix 5A: The data used in this chapter

The symbols used here have already been described in this chapter, where i=1,...,m, refer to the number of blocks and j=1,...,k, the number of observations in the ith block.

Table 5.1
Australian cross-sectional data for 1961 and 1976 for n=96, m=12 and k=8

i	j	x_{ij2}	x_{ij3}	x_{ij4}	i	j	x_{ij2}	x_{ij3}	x_{ij4}
1	1	4.12100	0.02555	0.0062	4	1	0.00020	0.00001	0.0500
1	2	2.63500	0.11010	0.0418	4	2	0.00149	0.00020	0.1342
1	3	1.87800	0.24480	0.1304	4	3	0.01317	0.00481	0.3652
1	4	0.89820	0.21520	0.2396	4	4	0.04521	0.02648	0.5857
1	5	0.63030	0.21200	0.3364	4	5	0.11530	0.07438	0.6452
1	6	0.24350	0.09219	0.3786	4	6	0.09287	0.05991	0.6451
1	7	0.20460	0.07880	0.3852	4	7	0.12980	0.07857	0.6054
1	8	0.38840	0.15510	0.3994	4	8	0.76280	0.39370	0.5161
2	1	0.03885	0.01566	0.4031	5	1	3.69400	0.01921	0.0052
2	2	0.97860	0.67680	0.6916	5	2	1.32600	0.04987	0.0376
2	3	5.35100	4.53400	0.8473	5	3	0.65880	0.05916	0.0898
2	4	6.34100	5.73600	0.9046	5	4	0.46330	0.07904	0.1706
2	5	5.42600	4.95500	0.9132	5	5	0.48050	0.13710	0.2853
2	6	2.00100	1.81800	0.9084	5	6	0.23300	0.08672	0.3722
2	7	1.54300	1.38800	0.8996	5	7	0.22960	0.09442	0.4113
2	8	2.64900	2.32200	0.8766	5	8	0.64360	0.27550	0.4281
3	1	0.00009	0.00000	0.0500	6	1	0.27610	0.00218	0.0079
3	2	0.00180	0.00016	0.0889	6	2	2.02800	0.02961	0.0416
3	3	0.04473	0.01209	0.2703	6	3	5.89300	0.15150	0.0257
3	4	0.10790	0.04122	0.3819	6	4	6.48800	0.23940	0.0369
3	5	0.11680	0.04937	0.4226	6	5	4.92400	0.23440	0.0476
3	6	0.04282	0.01870	0.4367	6	6	1.61200	0.08206	0.0509
3	7	0.03126	0.01354	0.4331	6	7	1.29000	0.07017	0.0544
3	8	0.04093	0.01796	0.4388	6	8	1.87800	0.13630	0.0726

Table 5.1 continued

i	j	x_{ij2}	x_{ij3}	x_{ij4}	i	j	x_{ij2}	x_{ij3}	x_{ij4}
7	1	0.00030	0.00003	0.1000	10	1	0.06212	0.04344	0.6671
7	2	0.00521	0.00068	0.1305	10	2	1.83700	1.55700	0.8475
7	3	0.06138	0.01899	0.3094	10	3	8.19900	7.53000	0.9184
7	4	0.12250	0.05923	0.4834	10	4	6.92300	6.56100	0.9477
7	5	0.12400	0.07017	0.5660	10	5	6.68500	6.34100	0.9486
7	6	0.04237	0.02350	0.5546	10	6	2.65000	2.49400	0.9411
7	7	0.03383	0.01829	0.5406	10	7	2.28300	2.12000	0.9290
7	8	0.04378	0.02294	0.5240	10	8	3.70000	3.32600	0.8989
8	1	0.00071	0.00009	0.1268	11	1	0.00055	0.00012	0.2182
8	2	0.00498	0.00148	0.2972	11	2	0.02006	0.00830	0.4138
8	3	0.03950	0.02376	0.6016	11	3	0.22330	0.13280	0.5949
8	4	0.15570	0.11420	0.7334	11	4	0.21490	0.13580	0.6321
8	5	0.43840	0.32400	0.7391	11	5	0.23890	0.15200	0.6363
8	6	0.36430	0.25490	0.6996	11	6	0.09308	0.05882	0.6319
8	7	0.54610	0.36350	0.6656	11	7	0.07435	0.04426	0.5953
8	8	2.53700	1.40600	0.5544	11	8	0.09513	0.05479	0.5759
9	1	6.15200	0.13720	0.0223	12	1	0.00061	0.00012	0.1967
9	2	3.72800	0.60800	0.1631	12	2	0.00257	0.00090	0.3502
9	3	2.08600	0.68540	0.3286	12	3	0.01539	0.00876	0.5692
9	4	0.71950	0.28880	0.4014	12	4	0.03801	0.02700	0.7103
9	5	0.65550	0.31160	0.4753	12	5	0.12640	0.09649	0.7634
9	6	0.23960	0.12600	0.5260	12	6	0.10500	0.07670	0.7308
9	7	0.20910	0.11500	0.5502	12	7	0.14970	0.10760	0.7191
9	8	0.41130	0.20610	0.5011	12	8	0.85720	0.50310	0.5869

Table 5.2
Australian cross-sectional data for 1966 and 1971 for n=96, m=2, k=48, i.e. i=1,2 and j=1,...,48

i	j	x_{ij2}	x_{ij3}	x_{ij4}	i	j	x_{ij2}	x_{ij3}	x_{ij4}
1	1	5.30000	0.03816	0.0072	1	37	0.41670	0.13410	0.3217
1	2	3.05300	0.16760	0.0549	1	38	0.22290	0.09266	0.4157
1	3	1.70300	0.27380	0.1608	1	39	0.21990	0.10010	0.4551
1	4	0.92600	0.25690	0.2774	1	40	0.67400	0.31150	0.4621
1	5	0.62020	0.23440	0.3779	1	41	0.41560	0.00424	0.0102
1	6	0.25580	0.10920	0.4269	1	42	2.47000	0.04841	0.0196
1	7	0.20670	0.09189	0.4445	1	43	6.16300	0.19410	0.0315
1	8	0.39730	0.17040	0.4290	1	44	6.73800	0.29240	0.0434
1	9	0.06853	0.03761	0.5488	1	45	5.48500	0.26440	0.0482
1	10	1.30500	0.99830	0.7652	1	46	1.96700	0.09618	0.0489
1	11	5.64200	4.93100	0.8739	1	47	1.38000	0.06762	0.0490
1	12	6.85500	6.32300	0.9224	1	48	2.02800	0.11660	0.0575
1	13	5.80200	5.39500	0.9298					
1	14	2.35100	2.17000	0.9232	2	1	0.00083	0.00003	0.0361
1	15	1.78000	1.62600	0.9134	2	2	0.00733	0.00163	0.2224
1	16	2.87000	2.54500	0.8867	2	3	0.06373	0.02638	0.4139
1	17	0.00102	0.00007	0.0686	2	4	0.12950	0.07305	0.5642
1	18	0.00373	0.00054	0.1448	2	5	0.14500	0.09018	0.6220
1	19	0.04243	0.01470	0.3464	2	6	0.05808	0.03590	0.6181
1	20	0.11010	0.04925	0.4474	2	7	0.04308	0.02592	0.6017
1	21	0.12740	0.06448	0.5061	2	8	0.06390	0.03633	0.5686
1	22	0.05360	0.02643	0.4931	2	9	0.00151	0.00015	0.0993
1	23	0.03912	0.02021	0.5166	2	10	0.00684	0.00230	0.3363
1	24	0.05147	0.02452	0.4764	2	11	0.03992	0.02643	0.6621
1	25	0.00100	0.00005	0.0500	2	12	0.16050	0.12990	0.8093
1	26	0.00335	0.00040	0.1194	2	13	0.47810	0.37260	0.7794
1	27	0.01274	0.00525	0.4121	2	14	0.41990	0.30570	0.7280
1	28	0.04883	0.03092	0.6332	2	15	0.55240	0.38200	0.6915
1	29	0.11860	0.08314	0.7012	2	16	2.97200	1.70500	0.5738
1	30	0.10050	0.06866	0.6833	2	17	5.58500	0.06143	0.0110
1	31	0.13000	0.08465	0.6514	2	18	3.56700	0.31710	0.0889
1	32	0.80640	0.42790	0.5306	2	19	1.80600	0.39720	0.2199
1	33	4.69900	0.03477	0.0074	2	20	0.82120	0.26850	0.3269
1	34	1.68200	0.10160	0.0604	2	21	0.63840	0.26860	0.4208
1	35	0.66970	0.09041	0.1350	2	22	0.24850	0.12100	0.4869
1	36	0.41280	0.08611	0.2086	2	23	0.20490	0.10350	0.5051

108

Table 5.2 continued

i	j	x_{ij2}	x_{ij3}	x_{ij4}	i	j	x_{ij2}	x_{ij3}	x_{ij4}
2	24	0.39510	0.19180	0.4855	2	37	0.17340	0.09697	0.5593
2	25	0.08158	0.04674	0.5729	2	38	0.07357	0.04250	0.5777
2	26	2.00000	1.62400	0.8120	2	39	0.05676	0.03222	0.5677
2	27	7.02100	6.31100	0.8989	2	40	0.07699	0.04222	0.5484
2	28	6.88000	6.43100	0.9347	2	41	0.00033	0.00009	0.2727
2	29	6.37700	5.99400	0.9399	2	42	0.00288	0.00094	0.3264
2	30	2.58500	2.41800	0.9354	2	43	0.01566	0.00844	0.5390
2	31	2.03200	1.88100	0.9257	2	44	0.05032	0.03571	0.7097
2	32	3.15600	2.82800	0.8959	2	45	0.13350	0.10110	0.7575
2	33	0.00000	0.00000	0.0000	2	46	0.10810	0.07992	0.7394
2	34	0.01401	0.00325	0.2320	2	47	0.14400	0.10200	0.7086
2	35	0.08930	0.03885	0.4351	2	48	0.84040	0.48120	0.5725
2	36	0.13350	0.06901	0.5169					

Table 5.3
Artificially generated log-normal variable data for n=96, m=3, k=32, i.e. i=1,2,3, j=1,...,32

i	j	x_{ij2}	i	j	x_{ij2}	i	j	x_{ij2}
1	1	1.33369	2	1	1.31581	3	1	0.96271
1	2	1.15899	2	2	1.41961	3	2	0.97292
1	3	1.34683	2	3	1.37922	3	3	0.81020
1	4	0.82016	2	4	1.29287	3	4	0.88585
1	5	1.21295	2	5	1.06239	3	5	1.11333
1	6	1.03895	2	6	1.63080	3	6	0.87408
1	7	1.13887	2	7	1.06660	3	7	1.39801
1	8	0.51691	2	8	1.08967	3	8	1.05341
1	9	1.38710	2	9	1.11439	3	9	1.30407
1	10	1.11005	2	10	0.74284	3	10	1.25518
1	11	0.31322	2	11	1.67184	3	11	1.32082
1	12	1.07261	2	12	1.61571	3	12	0.32865
1	13	0.64652	2	13	0.97981	3	13	0.49337
1	14	1.09548	2	14	1.07496	3	14	0.90605
1	15	1.35041	2	15	1.02444	3	15	1.22240
1	16	1.49629	2	16	1.56426	3	16	0.73985
1	17	1.36745	2	17	0.26251	3	17	0.64744
1	18	1.45277	2	18	1.28764	3	18	0.80285
1	19	1.29114	2	19	1.21660	3	19	0.79286
1	20	1.43091	2	20	0.65502	3	20	1.08637
1	21	0.30469	2	21	1.69976	3	21	1.48146
1	22	-0.21046	2	22	1.12795	3	22	1.34389
1	23	1.03297	2	23	1.46930	3	23	0.66093
1	24	1.10924	2	24	1.06172	3	24	1.26205
1	25	1.51238	2	25	1.38130	3	25	0.95833
1	26	1.17286	2	26	1.25050	3	26	1.42739
1	27	0.66592	2	27	-0.41641	3	27	1.22635
1	28	1.42282	2	28	0.95260	3	28	1.41553
1	29	1.34428	2	29	1.01722	3	29	1.09891
1	30	1.20788	2	30	0.70807	3	30	1.14698
1	31	1.16500	2	31	1.48471	3	31	0.81043
1	32	0.80592	2	32	1.46623	3	32	1.45789

APPENDIX 5B:
Tabulation of results of the comparative study

Table 5.4

Comparison of an exact and approximate critical values of $s(\rho_1)$ test for block of size $\dfrac{n}{2}$, p = 4, using cross-sectional data for 1961 and 1976

POI Test	ρ_1	Exact critical values, c_α at $\alpha=.05$	Approximate critical values, c_α^*	Estimated sizes based on c_α^*
		n=32, F(1,27)=4.2097		
$s(\rho_1)$	0.1	1.03417	0.96124	.005
	0.2	1.13137	1.08139	.018
	0.3	1.27380	1.23588	.027
	0.4	1.47252	1.44186	.034
	0.5	1.75592	1.73023	.039
		n=64, F(1,59)=4.003		
$s(\rho_1)$	0.1	1.05750	1.04051	.024
	0.2	1.18040	1.17058	.036
	0.3	1.34470	1.33780	.041
	0.4	1.56608	1.56077	.044
	0.5	1.87723	1.87293	.046
		n=96, F(1,91)=3.9454		
$s(\rho_1)$	0.1	1.07339	1.06494	.030
	0.2	1.20276	1.19806	.040
	0.3	1.37246	1.36921	.044
	0.4	1.59990	1.59741	.046
	0.5	1.91890	1.91689	.047

Table 5.5

Comparison of an exact and approximate critical values of $s(\rho_1)$ test for block of size $\frac{n}{2}$, p = 4, using cross-sectional data for 1966 and 1971

POI Test	ρ_1	Exact critical values, c_α at $\alpha=.05$	Approximate critical values, c_α^*	Estimated sizes based on c_α^*
		n=32, F(1,27)=4.2097		
$s(\rho_1)$	0.1	1.03296	0.96124	.005
	0.2	1.13024	1.08139	.018
	0.3	1.27291	1.23588	.028
	0.4	1.47167	1.44186	.034
	0.5	1.75518	1.73023	.039
		n=64, F(1,59)=4.003		
$s(\rho_1)$	0.1	1.05755	1.04051	.024
	0.2	1.18040	1.17058	.036
	0.3	1.34470	1.33780	.041
	0.4	1.56608	1.56077	.044
	0.5	1.87724	1.87293	.046
		n=96, F(1,91)=3.9454		
$s(\rho_1)$	0.1	1.07293	1.06494	.031
	0.2	1.20248	1.19806	.040
	0.3	1.37226	1.36921	.044
	0.4	1.59974	1.59741	.046
	0.5	1.91877	1.91689	.047

112

Table 5.6

Comparison of an exact and approximate critical values of s(ρ_1) test for block of size $\frac{n}{2}$, p = 2, using artificially generated log-normal regressor

POI Test	ρ_1	Exact critical values, c_α at $\alpha=.05$	Approximate critical values, c_α^*	Estimated sizes based on c_α^*
		n=32, F(1,27)=4.2097		
s(ρ_1)	0.1	1.021471	0.96124	.009
	0.2	1.123943	1.08139	.022
	0.3	1.271409	1.23588	.029
	0.4	1.474579	1.44186	.033
	0.5	1.762715	1.73023	.036
		n=64, F(1,59)=4.003		
s(ρ_1)	0.1	1.057875	1.04051	.023
	0.2	1.181720	1.17058	.034
	0.3	1.134672	1.33780	.039
	0.4	1.568771	1.56077	.041
	0.5	1.880729	1.87293	.043
		n=96, F(1,91)=3.9454		
s(ρ_1)	0.1	1.073136	1.06494	.030
	0.2	1.203116	1.19806	.039
	0.3	1.373191	1.36921	.042
	0.4	1.600954	1.59741	.044
	0.5	1.920328	1.91689	.047

Table 5.7
Powers of the POI-tests for the $n \times 4$ design matrix of cross-sectional data for 1961 and 1976, n=32, p=4 and $\alpha = .05$

Tests	c_α	$\rho=0.1$	0.2	0.3	0.4	0.5	Average Power
			Block size 2				
s(.1)	0.97890	.0989	.1783	.2943	.4455	.6180	.3270
s(.2)	0.97657	.0989	.1783	.2944	.4458	.6186	.3272
s(.3)	0.99450	.0989	.1783	.2944	.4460	.6191	.3273
s(.4)	1.03719	.0989	.1782	.2944	.4461	.6196	.3274
s(.5)	1.11435	.0988	.1780	.2941	.4460	.6198	.3273
			Block size 4				
s(.1)	0.98451	.1397	.2720	.4271	.5831	.7230	.4290
s(.2)	1.01296	.1395	.2724	.4290	.5868	.7282	.4312
s(.3)	1.07863	.1390	.2721	.4295	.5889	.7316	.4322
s(.4)	1.18630	.1383	.2710	.4290	.5895	.7335	.4323
s(.5)	1.35232	.1373	.2692	.4274	.5889	.7341	.4314
			Block size 8				
s(.1)	1.00513	.1640	.3016	.4375	.5626	.6740	.4279
s(.2)	1.06902	.1637	.3026	.4410	.5684	.6812	.4314
s(.3)	1.17651	.1627	.3021	.4418	.5705	.6843	.4323
s(.4)	1.33503	.1615	.3009	.4413	.5710	.6854	.4320
s(.5)	1.56785	.1604	.2995	.4404	.5707	.6857	.4313
			Block size 16				
s(.1)	1.03417	.1638	.2738	.3702	.4551	.5318	.3589
s(.2)	1.13137	.1638	.2738	.3702	.4551	.5318	.3589
s(.3)	1.27380	.1638	.2738	.3702	.4551	.5318	.3589
s(.4)	1.47252	.1638	.2738	.3702	.4551	.5318	.3589
s(5)	1.75592	.1638	.2738	.3702	.4551	.5318	.3589

Table 5.8
Powers of the POI-tests for the n×4 design matrix of cross-sectional data
for 1966 and 1971, n=32, p=4 and α=.05

Tests	c_α	ρ=0.1	0.2	0.3	0.4	0.5	Average Power
			Block size 2				
s(.1)	0.97901	.0990	.1785	.2947	.4461	.6186	.3274
s(.2)	0.97683	.0990	.1785	.2947	.4463	.6192	.3275
s(.3)	0.99496	.0990	.1785	.2948	.4465	.6197	.3277
s(.4)	1.03793	.0989	.1784	.2947	.4466	.6203	.3279
s(.5)	1.11546	.0988	.1782	.2945	.4465	.6203	.3277
			Block size 4				
s(.1)	0.98297	.1402	.2738	.4306	.5883	.7292	.4324
s(.2)	1.01008	.1400	.2742	.4326	.5921	.7343	.4346
s(.3)	1.07452	.1395	.2738	.4331	.5941	.7376	.4356
s(.4)	1.18106	.1388	.2728	.4326	.5947	.7393	.4356
s(.5)	1.34601	.1379	.2712	.4311	.5941	.7399	.4348
			Block size 8				
s(.1)	1.00358	.1681	.3084	.4450	.5691	.6786	.4338
s(.2)	1.06773	.1676	.3094	.4484	.5750	.6861	.4373
s(.3)	1.17569	.1665	.3088	.4493	.5773	.6895	.4383
s(.4)	1.33462	.1653	.3076	.4488	.5778	.6908	.4381
s(.5)	1.56777	.1640	.3060	.4478	.5775	.6911	.4373
			Block size16				
s(.1)	1.03296	.1672	.2792	.3763	.4613	.5378	.3644
s(.2)	1.13024	.1672	.2792	.3763	.4613	.5378	.3644
s(.3)	1.27291	.1672	.2792	.3763	.4613	.5377	.3643
s(.4)	1.47167	.1672	.2792	.3763	.4613	.5378	.3644
s(5)	1.75518	.1672	.2792	.3763	.4613	.5378	.3644

Table 5.9
Powers of the POI-tests for the n×2 design matrix of the artificially generated log-normal regressor, n=32, p=4 and α=.05

Tests	c_α	$\rho=0.1$	0.2	0.3	0.4	0.5	Average Power
			Block size 2				
s(.1)	0.97330	.1009	.1849	.3089	.4705	.6517	.3434
s(.2)	0.96527	.1009	.1849	.3090	.4709	.6525	.3436
s(.3)	0.97711	.1009	.1849	.3091	.4712	.6531	.3438
s(.4)	1.01304	.1008	.1848	.3090	.4713	.6536	.3439
s(.5)	1.08243	.1007	.1846	.3088	.4711	.6538	.3438
			Block size 4				
s(.1)	0.97022	.1562	.3188	.5046	.6783	.8167	.4949
s(.2)	0.99060	.1561	.3189	.5053	.6794	.8179	.4955
s(.3)	1.05195	.1560	.3188	.5054	.6798	.8184	.4957
s(.4)	1.15833	.1558	.3186	.5053	.6799	.8186	.4956
s(.5)	1.32636	.1556	.3183	.5051	.6798	.8187	.4955
			Block size 8				
s(.1)	0.98236	.2185	.4109	.5757	.7048	.8024	.5425
s(.2)	1.04442	.2185	.4109	.5758	.7048	.8024	.5425
s(.3)	1.15544	.2185	.4109	.5758	.7048	.8024	.5425
s(.4)	1.31982	.2185	.4109	.5758	.7048	.8024	.5425
s(.5)	1.56046	.2185	.4109	.5758	.7048	.8024	.5425
			Block size 16				
s(.1)	1.02147	.2296	.3679	.4715	.5540	.6236	.4493
s(.2)	1.12394	.2296	.3679	.4715	.5540	.6236	.4493
s(.3)	1.27141	.2296	.3679	.4715	.5540	.6236	.4493
s(.4)	1.47458	.2296	.3679	.4715	.5540	.6236	.4493
s(.5)	1.76272	.2296	.3679	.4715	.5540	.6236	.4493

Table 5.10
Powers of the POI-tests for the nx4 design matrix of cross-sectional data
for 1961 and 1976, n=64, p=4, and α=.05

Tests	c_α	$\rho=0.1$	0.2	0.3	0.4	0.5	Average Power
			Block size 2				
s(.1)	0.98578	.1337	.2878	.5048	.7319	.8990	.5114
s(.2)	0.99114	.1337	.2878	.5049	.7321	.8992	.5115
s(.3)	1.01822	.1337	.2878	.5049	.7323	.8995	.5116
s(.4)	1.07265	.1336	.2877	.5049	.7323	.8996	.5116
s(.5)	1.16612	.1336	.2877	.5049	.7322	.8997	.5116
			Block size 4				
s(.1)	0.98972	.2198	.4845	.7301	.8887	.9647	.6576
s(.2)	1.02571	.2197	.4848	.7311	.8898	.9654	.6582
s(.3)	1.10161	.2193	.4845	.7314	.8904	.9659	.6583
s(.4)	1.22337	.2187	.4838	.7312	.8906	.9661	.6581
s(.5)	1.40963	.2180	.4826	.7305	.8904	.9662	.6575
			Block size 8				
s(.1)	1.00735	.2968	.5737	.7696	.8860	.9488	.6950
s(.2)	1.07986	.2962	.5745	.7719	.8884	.9506	.6964
s(.3)	1.19791	.2949	.5742	.7724	.8892	.9513	.6964
s(.4)	1.36903	.2934	.5730	.7721	.8894	.9516	.6959
s(.5)	1.61787	.2919	.5716	.7714	.8893	.9517	.6952
			Block size 16				
s(.1)	1.02967	.3337	.5705	.7247	.8242	.8898	.6686
s(.2)	1.13044	.3341	.5708	.7252	.8248	.8902	.6690
s(.3)	1.27632	.3331	.5707	.7253	.8249	.8902	.6689
s(.4)	1.47822	.3328	.5706	.7253	.8249	.8903	.6688
s(.5)	1.76525	.3327	.5705	.7252	.8249	.8903	.6687
			Block size 32				
s(.1)	1.05750	.3297	.4841	.5829	.6552	.7130	.5530
s(.2)	1.18040	.3297	.4841	.5829	.6552	.7130	.5530
s(.3)	1.34470	.3297	.4841	.5829	.6552	.7130	.5530
s(.4)	1.56608	.3297	.4841	.5829	.6552	.7130	.5530
s(.5)	1.87723	.3297	.4841	.5829	.6552	.7130	.5530

117

Table 5.11
Powers of the POI-tests for the nx4 design matrix of cross-sectional data
for 1966 and 1971, n=64, p=4 and α=.05

Tests	c_α	$\rho=0.1$	0.2	0.3	0.4	0.5	Average Power
			Block size 2				
s(.1)	0.98576	.1337	.2877	.5046	.7317	.8989	.5113
s(.2)	0.99108	.1337	.2877	.5047	.7319	.8991	.5114
s(.3)	1.01812	.1337	.2877	.5048	.7321	.8994	.5115
s(.4)	1.07249	.1336	.2876	.5047	.7322	.8995	.5115
s(.5)	1.16585	.1335	.2874	.5045	.7321	.8996	.5114
			Block size 4				
s(.1)	0.98940	.2203	.4855	.7314	.8896	.9651	.6589
s(.2)	1.02518	.2201	.4858	.7323	.8907	.9658	.6589
s(.3)	1.10092	.2197	.4856	.7326	.8913	.9663	.6591
s(.4)	1.22252	.2192	.4849	.7324	.8914	.9665	.6589
s(.5)	1.40862	.2184	.4837	.7317	.8913	.9666	.6583
			Block size 8				
s(.1)	1.00715	.2982	.5755	.7709	.8867	.9491	.6961
s(.2)	1.07972	.2976	.5765	.7732	.8891	.9509	.6975
s(.3)	1.19786	.2963	.5760	.7737	.8900	.9516	.6975
s(.4)	1.36905	.2948	.5748	.7734	.8901	.9520	.6970
s(.5)	1.61794	.2932	.5733	.7727	.8900	.9520	.6962
			Block size 16				
s(.1)	1.02978	.3329	.5695	.7238	.8236	.8893	.6678
s(.2)	1.13054	.3326	.5698	.7244	.8241	.8897	.6681
s(.3)	1.27641	.3323	.5698	.7244	.8242	.8898	.6681
s(.4)	1.47829	.3321	.5696	.7244	.8242	.8899	.6680
s(.5)	1.76531	.3319	.5695	.7244	.8242	.8899	.6680
			Block size 32				
s(.1)	1.05755	.3295	.4840	.5827	.6550	.7129	.5528
s(.2)	1.18040	.3295	.4840	.5827	.6550	.7129	.5528
s(.3)	1.34470	.3295	.4840	.5827	.6550	.7129	.5528
s(.4)	1.56608	.3295	.4840	.5827	.6550	.7129	.5528
s(.5)	1.87724	.3295	.4840	.5827	.6550	.7129	.5528

Table 5.12
Powers of the POI-tests for the n×2 design matrix of the artificially
generated log-normal regressor, n=64, p=2 and α=.05

Tests	c_α	$\rho=0.1$	0.2	0.3	0.4	0.5	Average power
			Block size 2				
s(.1)	0.98283	.1357	.2937	.5166	.7465	.9100	.5204
s(.2)	0.98511	.1354	.2938	.5167	.7468	.9103	.5206
s(.3)	1.00889	.1354	.2937	.5168	.7470	.9105	.5207
s(.4)	1.05962	.1353	.2936	.5167	.7471	.9107	.5207
s(.5)	1.14882	.1352	.2934	.5165	.7470	.9108	.5206
			Block size 4				
s(.1)	0.98440	.2336	.5187	.7689	.9156	.9771	.6828
s(.2)	1.01766	.2336	.5188	.7691	.9158	.9772	.6829
s(.3)	1.09239	.2335	.5187	.7692	.9158	.9773	.6829
s(.4)	1.21420	.2334	.5186	7692	.9159	.9773	.6829
s(.5)	1.40184	.2333	.5185	.7691	.9159	.9773	.6828
			Block size 8				
s(.1)	0.99804	.3408	.6461	.8333	.9283	.9719	.7441
s(.2)	1.06948	.3408	.6461	.8333	.9283	.9719	.7441
s(.3)	1.18875	.3408	.6461	.8333	.9283	.9719	.7441
s(.4)	1.36214	.3408	.6461	.8333	.9283	.9719	.7441
s(.5)	1.61410	.3408	.6461	.8334	.9283	.9719	.7441
			Block size 16				
s(.1)	1.02404	.3979	.6444	.7850	.8682	.9198	.7231
s(.2)	1.12727	.3979	.6444	.7850	.8682	.9198	.7231
s(.3)	1.27539	.3979	.6444	.7850	.8682	.9198	.7231
s(.4)	1.47933	.3979	.6444	.7850	.8682	.9198	.7231
s(.5)	1.76849	.3979	.6444	.7850	.8682	.9198	.7231
			Block size 32				
s(.1)	1.05787	.3508	.5058	.6025	.6723	.7278	.5718
s(.2)	1.18172	.3508	.5058	.6025	.6723	.7129	.5718
s(.3)	1.34672	.3508	.5058	.6025	.6723	.7129	.5718
s(.4)	1.56877	.3508	.5058	.6025	.6723	.7129	.5718
s(.5)	1.88073	.3508	.5058	.6025	.6723	.7129	.5718

119

Table 5.13
Powers of the POI-tests for the n×4 design matrix of cross-sectional data for 1961 and 1976, n=96, p=4 and α=.05

Tests	c_α	$\rho=0.1$	0.2	0.3	0.4	0.5	Average power
			Block size 2				
s(.1)	0.98950	.1647	.3858	.6639	.8808	.9771	.6145
s(.2)	0.99894	.1646	.3859	.6640	.8809	.9772	.6145
s(.3)	1.03081	.1646	.3858	.6640	.8810	.9773	.6145
s(.4)	1.09127	.1646	.3858	.6640	.8810	.9773	.6145
s(.5)	1.19295	.1645	.3856	.6638	.8810	.9773	.6144
			Block size 4				
s(.1)	0.99500	.2887	.6389	.8779	.9723	.9959	.7547
s(.2)	1.03622	.2886	.6391	.8784	.9726	.9960	.7549
s(.3)	1.11791	.2881	.6387	.8785	.9727	.9960	.7548
s(.4)	1.24677	.2877	.6384	.8784	.9728	.9961	.7547
s(.5)	1.44241	.2870	.6375	.8781	.9728	.9961	.7543
			Block size 8				
s(.1)	1.01165	.4056	.7466	.9119	.9734	.9931	.8061
s(.2)	1.08858	.4051	.7472	.9127	.9739	.9933	.8064
s(.3)	1.21144	.4040	.7469	.9129	.9742	.9934	.8063
s(.4)	1.38817	.4028	.7462	.9128	.9742	.9935	.8059
s(.5)	1.64415	.4014	.7452	.9125	.9742	.9935	.8054
			Block size 16				
s(.1)	1.03267	.4741	.7561	.8872	.9477	.9764	.8083
s(.2)	1.13678	.4738	.7564	.8875	.9480	.9765	.8084
s(.3)	1.28555	.4738	.7563	.8876	.9481	.9766	.8085
s(.4)	1.49049	.4731	.7562	.8876	.9481	.9766	.8083
s(.5)	1.78123	.4728	.7561	.8875	.9481	.9766	.8082
			Block size 48				
s(.1)	1.07339	.3977	.5515	.6428	.7072	.7576	.6114
s(.2)	1.20276	.3977	.5515	.6428	.7072	.7576	.6114
s(.3)	1.37246	.3977	.5515	.6428	.7072	.7576	.6114
s(.4)	1.59990	.3977	.5515	.6428	.7072	.7576	.6114
s(.5)	1.91890	.3977	.5515	.6428	.7072	.7576	.6114

120

Table 5.14
Powers of the POI-tests for the nx4 design matrix of cross-sectional data
for 1966 and 1971, n=96, p=4 and α=.05

Tests	c_α	$\rho=0.1$	0.2	0.3	0.4	0.5	Average power
			Block size 2				
s(.1)	0.98948	.1646	.3857	.6637	.8807	.9771	.6144
s(.2)	0.99888	.1646	.3858	.6638	.8808	.9771	.6144
s(.3)	1.03070	.1646	.3857	.6638	.8809	.9772	.6144
s(.4)	1.09110	.1646	.3856	.6638	.8809	.9773	.6143
s(.5)	1.19267	.1645	.3854	.6636	.8809	.9773	.6143
			Block size 4				
s(.1)	0.99479	.2891	.6396	.8785	.9725	.9960	.7551
s(.2)	1.03587	.2890	.6398	.8790	.9728	.9960	.7553
s(.3)	1.11746	.2886	.6396	.8791	.9730	.9961	.7553
s(.4)	1.24618	.2881	.6391	.8790	.9730	.9961	.7551
s(.5)	1.44169	.2874	.6382	.8786	.9730	.9961	.7547
			Block size 8				
s(.1)	1.01121	.4080	.7494	.9134	.9740	.9933	.8076
s(.2)	1.08807	.4075	.7500	.9143	.9746	.9935	.8080
s(.3)	1.21090	.4065	.7497	.9144	.9748	.9936	.8078
s(.4)	1.38764	.4052	.7489	.9143	.9748	.9937	.8074
s(.5)	1.64361	.4040	.7481	.9141	.9748	.9937	.8069
			Block size 16				
s(.1)	1.03208	.4802	.7616	.8905	.9496	.9773	.8118
s(.2)	1.13627	.4799	.7618	.8908	.9498	.9774	.8119
s(.3)	1.28513	.4795	.7618	.8909	.9498	.9775	.8119
s(.4)	1.49013	.4792	.7617	.8909	.9498	.9775	.8118
s(.5)	1.78092	.4792	.7616	.8908	.9498	.9775	.8118
			Block size 48				
s(.1)	1.07293	.4107	.5636	.6533	.7162	.7652	.6218
s(.2)	1.20248	.4107	.5636	.6533	.7162	.7652	.6218
s(.3)	1.37226	.4107	.5636	.6533	.7162	.7652	.6218
s(.4)	1.59974	.4107	.5636	.6533	.7162	.7652	.6218
s(.5)	1.91877	.4107	.5636	.6533	.7162	.7652	.6218

Table 5.15
Powers of the POI-tests for the n×2 design matrix of artificially
generated log-normal regressor, n=96, p=2 and α=.05

Tests	c_α	$\rho=0.1$	0.2	0.3	0.4	0.5	Average power
			Block size 2				
s(.1)	0.98743	.1662	.3913	.6728	.8882	.9800	.6197
s(.2)	0.99470	.1662	.3913	.6729	.8883	.9801	.6198
s(.3)	1.02422	.1662	.3913	.6729	.8883	.9801	.6198
s(.4)	1.08204	.1661	.3912	.6728	.8884	.9801	.6197
s(.5)	1.18066	.1660	.3910	.6727	.8883	.9801	.6196
			Block size 4				
s(.1)	0.99157	.3016	.6646	.8968	.9794	.9974	.7680
s(.2)	1.03103	.3015	.6646	.8969	.9795	.9974	.7680
s(.3)	1.11199	.3015	.6646	.8969	.9795	.9974	.7680
s(.4)	1.24083	.3014	.6645	.8969	.9795	.9974	.7679
s(.5)	1.43734	.3013	.6044	.8969	.9795	.9974	.7679
			Block size 8				
s(.1)	1.00607	.4430	.7910	.9370	.9835	.9963	.8302
s(.2)	1.08236	.4430	.7910	.9370	.9835	.9963	.8302
s(.3)	1.20591	.4430	.7910	.9370	.9835	.9963	.8302
s(.4)	1.38397	.4430	.7910	.9370	.9835	.9963	.8302
s(.5)	1.64179	.4430	.7910	.9370	.9835	.9963	.8302
			Block size 16				
s(.1)	1.02952	.5252	.7998	.9131	.9615	.9831	.8365
s(.2)	1.13507	.5252	.7998	.9131	.9615	.9831	.8365
s(.3)	1.28517	.5252	.7998	.9131	.9615	.9831	.8365
s(.4)	1.49133	.5252	.7998	.9131	.9615	.9831	.8365
s(.5)	1.78334	.5252	.7998	.9131	.9615	.9831	.8365
			Block size 48				
s(.1)	1.07314	.4294	.5807	.6679	.7286	.7758	.6365
s(.2)	1.20312	.4294	.5807	.6679	.7286	.7758	.6365
s(.3)	1.37319	.4294	.5807	.6679	.7286	.7758	.6365
s(.4)	1.60095	.4294	.5807	.6679	.7286	.7758	.6365
s(.5)	1.92033	.4294	.5807	.6679	.7286	.7758	.6365

Table 5.16
Powers of the Durbin-Watson, one- and two-sided Lagrange multiplier and POI-tests for the nx4 design matrix of cross-sectional data for 1961 and 1976, n=32, p=4 and α=.05

Tests	$\rho=0.1$	0.2	0.3	0.4	0.5	Average power
			Block size 2			
PE	.099	.178	.294	.446	.620	-
s(.3)	.099	.178	.294	.446	.619	.3272
LM1	.099	.178	.294	.445	.617	.3266
LM2	.066	.112	.196	.324	.492	.2380
d^*	.081	.126	.186	.265	.363	.2042
d	.081	.125	.184	.260	.356	.2012
			Block size 4			
PE	.140	.272	.430	.590	.734	-
s(.3)	.139	.272	.430	.589	.732	.4324
LM1	.139	.271	.423	.577	.715	.4250
LM2	.094	.190	.323	.475	.627	.3418
d^*	.100	.175	.272	.390	.522	.2918
d	.099	.172	.268	.384	.516	.2878
			Block size 8			
PE	.164	.303	.442	.571	.686	-
s(.3)	.163	.302	.442	.571	.684	.4321
LM1	.163	.296	.427	.547	.656	.4178
LM2	.118	.224	.343	.463	.578	.3452
d^*	.095	.162	.246	.345	.454	.2604
d	.095	.161	.245	.343	.452	.2592

123

Table 5.17
Powers of the Durbin-Watson, one- and two-sided Lagrange multiplier and POI-tests for the nx4 design matrix of cross-sectional data for 1966 and 1971, n=32, p=4 and $\alpha=.05$

Tests	$\rho=0.1$	0.2	0.3	0.4	0.5	Average power
			Block size 2			
PE	.099	.179	.295	.447	.620	-
s(.3)	.099	.179	.295	.447	.620	.3280
LM1	.099	.178	.295	.446	.618	.3074
LM2	.066	.112	.196	.324	.493	.2250
d^*	.082	.126	.187	.266	.365	.2052
d	.081	.125	.185	.262	.359	.1862
			Block size 4			
PE	.140	.274	.433	.595	740	-
s(.3)	.140	.274	.433	.594	.738	.4358
LM1	.140	.272	.427	.582	.721	.4284
LM2	.095	.191	.326	.480	.632	.3448
d^*	.102	.178	.278	.398	.533	.2978
d	.101	.175	.274	.394	.528	.2944
			Block size 8			
PE	.168	.309	.449	.579	.691	-
s(.3)	.167	.309	.449	.577	.690	.4384
LM1	.167	.303	.435	.555	.661	.4242
LM2	.121	.230	.352	.472	.585	.3520
d^*	.098	.167	.255	.357	.468	.2690
d	.097	.166	.254	.356	.466	.2678

Table 5.18
Powers of the Durbin-Watson, one- and two-sided Lagrange multiplier and POI-tests for the nx4 design matrix of cross-sectional data for 1961 and 1976, n=64, p=4 and α=.05

Tests	$\rho=0.1$	0.2	0.3	0.4	0.5	Average power
			Block size 2			
PE	.134	.288	.505	.732	.900	-
s(.3)	.134	.288	.505	.732	.899	.5116
LM1	.134	.288	.505	.732	.899	.5116
LM2	.084	.191	.379	.618	.832	.4208
d^*	.102	.186	.307	.460	.630	.3370
d	.102	.185	.305	.456	.624	.3344
			Block size 4			
PE	.220	.485	.731	.891	.966	-
s(.3)	.219	.485	.731	.890	.966	.6582
LM1	.220	.483	.728	.887	.963	.6562
LM2	.146	.373	.632	.830	.940	.5842
d^*	.143	.298	.493	.687	.842	.4926
d	.142	.295	.489	.683	.839	.4896
			Block size 8			
PE	.297	.575	.772	.889	.952	-
s(.3)	.295	.574	.772	.889	.951	.6962
LM1	.296	.596	.763	.880	.944	.6958
LM2	.215	.473	.689	.833	.919	.6258
d^*	.148	.304	.486	.657	.797	.4784
d	.147	.303	.485	.657	.797	.4778

Table 5.19
Powers of the Durbin-Watson, one- and two-sided Lagrange multiplier and POI-tests for the n×4 design matrix of cross-sectional data for 1966 and 1971, n=64, p=4 and α=.05

Tests	$\rho=0.1$	0.2	0.3	0.4	0.5	Average power
			Block size 2			
PE	.134	.288	.505	.732	.900	-
s(.3)	.134	.288	.505	.732	.899	.5116
LM1	.134	.288	.505	.731	.899	.5114
LM2	.084	.191	.379	.618	.832	.4208
d^*	.102	.186	.307	.460	.630	.3370
d	.102	.185	.305	.456	.624	.3344
			Block size 4			
PE	.220	.486	.733	.891	.967	-
s(.3)	.220	.486	.733	.890	.966	.6592
LM1	.220	.484	.729	.888	.964	.6570
LM2	.147	.374	.634	.831	.940	.5852
d^*	.144	.299	.495	.689	.844	.4942
d	.143	.297	.492	.686	.841	.4918
			Block size 8			
PE	.298	.577	.774	.890	.952	-
s(.3)	.296	.576	.774	.890	.952	.6976
LM1	.297	.571	.764	.881	.945	.6916
LM2	.216	.476	.691	.834	.920	.6274
d^*	.149	.306	.489	.661	.800	.4810
d	.149	.306	.489	.661	.801	.4812

Table 5.20
Powers of the Durbin-Watson, one- and two-sided Lagrange multiplier and POI-tests for the n×4 design matrix of cross-sectional data for 1961 and 1976, n=96, p=4 and α=.05

Tests	$\rho=0.1$	0.2	0.3	0.4	0.5	Average power
			Block size 2			
PE	.165	.386	.664	.881	.977	-
s(.3)	.165	.386	.664	.881	.977	.6146
LM1	.165	.386	.664	.881	.977	.6146
LM2	.103	.272	.542	.806	.954	.5354
d^*	.120	.242	.415	.616	.800	.4386
d	.120	.240	.412	.612	.796	.4360
			Block size 4			
PE	.289	.639	.879	.973	.996	-
s(.3)	.288	.639	.879	.973	.996	.7550
LM1	.289	.638	.877	.972	.996	.7544
LM2	.198	.528	.814	.951	.992	.6966
d^*	.178	.398	.648	.843	.950	.6034
d	.177	.396	.645	.840	.949	.6014
			Block size 8			
PE	.406	.747	.913	.974	.993	-
s(.3)	.404	.747	.913	.974	.993	.8062
LM1	.404	.744	.909	.972	.992	.8042
LM2	.306	.660	.868	.956	.988	7556
d^*	.194	.429	.664	.836	.935	.6116
d	.193	.427	.663	.835	.934	.6104

Table 5.21
Powers of the Durbin-Watson, one- and two-sided Lagrange multiplier and POI-tests for the nx4 design matrix of cross-sectional data for 1966 and 1971, n=96, p=4 and α=.05

Tests	$\rho=0.1$	0.2	0.3	0.4	0.5	Average power
			Block size 2			
PE	.165	.386	.664	.881	.977	-
s(.3)	.165	.386	.664	.881	.977	.6146
LM1	.165	.386	.664	.881	.977	.6146
LM2	.103	.272	.542	.806	.954	.5354
d^*	.120	.242	.415	.616	.800	.4386
d	.120	.241	.412	.612	.796	.4362
			Block size 4			
PE	.289	.640	.879	.973	.996	-
s(.3)	.288	.640	.879	.973	.996	.7554
LM1	.289	.639	.877	.972	.996	.7546
LM2	.199	.529	.815	.951	.992	.6972
d^*	.178	.400	.650	.845	.951	.6048
d	.177	.398	.647	.843	.950	.6030
			Block size 8			
PE	.408	.750	.914	.975	.994	-
s(.3)	.407	.750	.914	.975	.994	.8080
LM1	.407	.746	.911	.973	.993	.8060
LM2	.309	.664	.870	.957	.988	7576
d^*	.196	.434	.670	.840	.937	.6154
d	.196	.433	.669	.839	.936	.6146

6 Testing for block effects in multi-stage linear regression models

Introduction [1]

The previous chapter considered a linear regression model based on a two-stage clustered sampling design and compared the power of the POI test with that of the existing tests and the PE, for testing equicorrelation within blocks. This chapter extends the 2SLR model to more general situations by dividing each block further into subblocks and hence, it has been named as a 3SLR model in chapter two. When the procedure of further divisions of subblocks continues, it leads to a multistage model. The correlated errors within blocks and subblocks give rise to intra-block equicorrelation, ρ_1, and intra-subblock equicorrelation, ρ_2, respectively. The aim of this chapter is to derive some optimal tests for the twin problems of testing nonzero values of the intra-block equicorrelation, ρ_1, and intra-subblock equicorrelation, ρ_2, as follows:

1   H_0: $\rho_1 = \rho_2 = 0$, against H_a: $\rho_1 > 0$, $\rho_2 > 0$.

and

2   H_0: $\rho_1 = 0$, $\rho_2 > 0$, against H_a: $\rho_1 > 0$, $\rho_2 > 0$.

For the first problem the LMMPI and POI tests are derived, whereas for the second problem the exact POI test and the asymptotic one-sided and two-sided Lagrange multiplier (LM1 and LM2) tests are derived. The main objectives of this study are to assess the accuracy of the asymptotic critical values and to compare the powers of the asymptotic tests with those of the POI tests. This chapter also extends the three-stage model to the multi-stage linear regression model and deals with further complicated testing problems.

This chapter is structured as follows. In the ensuing section, the three-stage model is introduced and the LMMPI and POI tests are constructed for testing ρ_1 and ρ_2. The POI test is used to obtain the power envelope, which is used as the benchmark in assessing the small sample power performance of the LMMPI test. An empirical power comparison of the LMMPI test with that of the POI test is also conducted. In the following section, the problem of testing the null hypothesis of a zero intra-subblock equicorrelation coefficient, ρ_2, in the presence of a intrablock equicorrelation coefficient, ρ_1 is considered. Next, the testing procedure developed in the previous section is applied to selected design matrices in the section titled 'An empirical comparison of sizes and powers' to answer the following questions.

1   Do the POI tests when applied to small-sample cases have superior sizes and power properties compared to those of asymptotic tests?

2   How well do the LM tests perform compared to the POI tests?

In the second last section of this chapter the 3SLR model is extended to the multi-stage model and it generalizes the optimal testing procedures for testing the multi-block effects. The last section contains some final remarks.

Three-stage linear regression model and the tests

The model

This section extends the 2SLR model considered in chapter five to the situation where data is obtained from a three-stage cluster sampling design. Following the summary of the 3SLR model (2.5) of chapter two, it is supposed that the total of n observations are sampled from m first-stage blocks (or clusters), with m(i) second-stage subblocks from the ith block and with m(i,j) third stage observations from the jth subblock of the ith block, such that $n = \sum\limits_{i=1}^{m} \sum\limits_{j=1}^{m(i)} m(i,j)$. Thus, the three-stage linear regression model can be expressed as

$$y_{ijk} = \sum_{\ell=1}^{p} \beta_\ell x_{ijk\ell} + u_{ijk} \tag{6.1}$$

for observations $k = 1,2,...,m(i,j)$ from the jth subblock, subblocks, $j = 1,2,...,m(i)$, from the ith block, $i = 1,2,...,m$, with y_{ijk} as the dependent variable, and p independent variables $x_{ijk\ell}$, $\ell = 1,2,...,p$, one of which may be a constant. The decomposition of the error term u_{ijk} can be written as

$$u_{ijk} = v_i + v_{ij} + v_{ijk}$$

where v_i is the ith block effect, v_{ij} is the jth subblock effect in the ith block and v_{ijk} is the remaining random effect. These three components of u_{ijk} are assumed to be mutually independent and normally distributed with

$$E\left(v_i\right) = E\left(v_{ij}\right) = E\left(v_{ijk}\right) = 0 \tag{6.2}$$

and

$$\text{var}(v_i) = \sigma_1^2, \quad \text{var}\left(v_{ij}\right) = \sigma_2^2, \quad \text{var}\left(v_{ijk}\right) = \sigma_3^2$$

so that $E(u_{ijk}) = 0$ and

$$E\left(u_{ijk}u_{rst}\right) = \begin{cases} 0 & \text{for } i \neq r \quad \text{any } j,s,k,t \\ \sigma_1^2 & \text{for } i = r, \ j \neq s, \ \text{any } k,t \\ \sigma_1^2 + \sigma_2^2 & \text{for } i = r, \ j = s, \ k \neq t \\ \sigma_1^2 + \sigma_2^2 + \sigma_3^2 & \text{for } i = r, \ j = s, \ k = t \end{cases} \tag{6.3}$$

for $k,t=1,2,...,m(i,j)$,

$j,s=1,2,...,m(i)$

$i,r=1,2,...,m.$

This gives rise to intra-block equicorrelation, whose coefficient is

$$\rho_1 = \frac{\sigma_1^2}{\sigma^2} \tag{6.4}$$

and intra-subblock equicorrelation, with the coefficient

$$\rho_2 = \frac{\sigma_1^2}{\sigma^2} \tag{6.5}$$

where $\sigma^2 = \sigma_1^2 + \sigma_2^2 + \sigma_3^2$, so $0 \leq \rho_1, \rho_2 \leq 1$, $\rho_1 + \rho_2 \leq 1$. Obviously, ρ_1 and ρ_2 originate from v_i and v_{ij} in error term u_{ijk}, respectively.

The regression model (6.1), under (6.2), (6.3), (6.4) and (6.5), can be written more compactly in matrix form as[2]

$$y = X\beta + u \tag{6.6}$$

in which y and u are $n \times 1$, X is $n \times p$, ß is $p \times 1$ and

$$u \sim N\left(0, \sigma^2 \Omega(\rho_1, \rho_2)\right) \tag{6.7}$$

where $\Omega(\rho_1, \rho_2) = \overset{m}{\underset{i=1}{\oplus}} \Omega_i(\rho_1, \rho_2)$ is block diagonal, with submatrices defined by (3.26) or by (3.27).

Although calculations can be proceeded with unbalanced data,[3] for the convenience of this discussion in relation with chapters three and four, the remainder of this chapter is constrained to the situations where data are balanced over blocks and even over subblocks. So let $T = m(i,j)$, and further assume $s = m(i)$, $i = 1,...,m$, $j = 1,...,s$, then $\Omega(\rho_1, \rho_2)$ in (6.7) can be simplified in terms of the usual notation of D_1 and D_2 as follows:

$$\Omega(\rho_1, \rho_2) = (1 - \rho - \rho_2)I_n + \rho_1 D_1 + \rho_2 D_2 \tag{6.8}$$

where D_1 and D_2 are as in (3.31) and I_n is $n \times n$ identity matrix, such that $n = msT$. Now, m is the number of main blocks, s is the number of subblocks in each block and T stands for the number of observations in each subblock of the main block.

The error covariance matrix in (6.7) involves two unknown parameters, ρ_1 and ρ_2. The main problem of interest of this subsection is to test

$$H_0 : \rho_1 = \rho_2 = 0 \tag{6.9}$$

against

$$H_a : \rho_1 \geq 0, \ \rho_2 \geq 0, \ (\text{excluding } H_0). \tag{6.10}$$

This testing problem is invariant with respect to transformations of the form

$$y \rightarrow r_0 y + Xr, \tag{6.11}$$

where r_0 is a positive scalar and r is a p×1 vector. The m×1 vector

$$\vartheta = \frac{Py}{(y'P'Py')^{1/2}} \tag{6.12}$$

is a maximal invariant under this group of transformation, where $m' = n-p$, $M_x = I_n - X(X'X)^{-1}X'$ and P is an m'×n matrix such that $PP' = I_{m'}$ and $P'P = M_x$. The probability density function of ϑ under (6.7) and (6.8) can be shown to be

$$f(\vartheta, \rho_1, \rho_2) - 1/2 \Gamma\left(\frac{m'}{2}\right) \pi^{-m'/2} \left| P\Omega(\rho_1, \rho_2)P' \right|^{-1/2} \tag{6.13}$$
$$\left(\vartheta' \left(P\Omega(\rho_1, \rho_2)P' \right)^{-1} \vartheta \right)^{-m'/2} d\vartheta,$$

where $d\vartheta$ denotes the uniform measure on the unit sphere. Note that the only unknown parameters in (6.13) are ρ_1 and ρ_2.

Based on ϑ, when more than one parameter is being tested, King and Wu (1990) showed that a locally most mean powerful test for

$$H_0' : \theta = 0$$

against

$$H_a' : \theta_1 \geq 0, ..., \ \theta_p \geq 0_p, \ \theta \neq 0,$$

133

is to reject H_0' for small values of

$$d = \frac{\hat{u}'A\hat{u}}{\hat{u}'\hat{u}}, \qquad (6.14)$$

where θ is $p \times 1$ vector of parameters,

$$A = \sum_{i=1}^{p} A_i,$$

$$A_i = -\frac{\partial \Sigma(\theta)}{\partial \theta_i}\bigg|_{\theta=0}, \quad i = 1,\ldots,p, \qquad (6.15)$$

and $\hat{u}$ is a vector of OLS residuals from (6.7). When $p=1$, d in (6.14) reduces to the locally best invariant (LBI) test statistic proposed by King and Hillier (1985). For the problem of testing (6.9) against (6.10), the LMMPI test is determined by (6.14), with $A = A_1 + A_2$, where

$$A_1 = \bigoplus_{i=1}^{m} A_{1i} = \left(I_n - D_1\right)$$

$$A_2 = \bigoplus_{i=1}^{m} A_{2i} = \left(I_n - D_2\right)$$

and by (3.31), A can be simplified as,

$$A = \left(2I_n - D_1 - D_2\right).$$

Therefore d in (6.14), can be expressed as,

$$d = \frac{\hat{u}'A\hat{u}}{\hat{u}'\hat{u}} = \frac{\hat{u}'\left(2I_n - D_1 - D_2\right)\hat{u}}{\hat{u}'\hat{u}}$$

$$= 2 - \left[\left(\hat{u}'D_1\hat{u} + \hat{u}'D_2\hat{u}\right)/\hat{u}'\hat{u}\right]$$

$$= 2 - \sum_{i=1}^{m}\left(\sum_{j=1}^{s}\sum_{k=1}^{T}\hat{u}_{ijk}\right)^2 / \sum_{i=1}^{m}\sum_{j=1}^{s}\sum_{k=1}^{T}\hat{u}_{ijk}^2$$

134

$$-\sum_{i=1}^{m}\sum_{j=1}^{s}\left(\sum_{k=1}^{T}\hat{u}_{ijk}\right)^2 \Big/ \sum_{i=1}^{m}\sum_{j=1}^{s}\sum_{k=1}^{T}\hat{u}_{ijk}^2 \, . \tag{6.16}$$

Note that d in (6.16) is a ratio of quadratic forms in normal variates, so its critical values can be evaluated by using standard numerical techniques mentioned in earlier chapters. Further note that if T = 0, then d in (6.16) becomes d´ in (5.15). This implies that the data does not possess the subblock effects and hence one would prefer to use d´ test in (5.15) of chapter five.

Point optimal invariant test

For the testing problem in the previous subsection, King's (1987b) POI test can be used to obtain the maximum attainable power, i.e. the power envelope over the alternative hypothesis parameter space. For any particular test, its performance can be assessed by checking how close its power is to the PE. Here the POI test statistic is introduced. It will be used in the next subsection in the comparative power study to assess the power performance of the LMMPI test, which was derived in the previous subsection. As it has been noticed in the previous subsection that this testing problem is invariant under the group of transformation (6.11), with the maximal invariant vector, ϑ, given by (6.12) and the density of ϑ, by (6.13).

Suppose $(\rho_1,\rho_2)' = (\rho_1^*,\rho_2^*)'$ is a chosen point in the alternative parameter space. Based on the maximal invariant in (6.12), King's (1987b) equation (18) gives the POI test for testing (6.9) against (6.10), that is, to reject the null hypothesis for small values of

$$s = \frac{\hat{u}'\Delta\hat{u}}{\hat{u}'\hat{u}} \tag{6.17}$$

where

$$\Delta = \Omega^{*-1} - \Omega^{*-1}X(X'\Omega^{*-1}X)^{-1}X'\Omega^{*-1},$$

$$\Omega^* = \Omega(\rho_1^*,\rho_2^*)$$

135

and û again is the vector of OLS residuals, i.e. $\hat{u} = M_X y$. Thus, tests based on s give the maximum attainable power at $(\rho_1^*, \rho_2^*)'$ in the class of invariant tests.

Note that s in (6.17) is a ratio of quadratic forms in normal variates, so their critical values can be evaluated by Imhof's algorithm, or Shively, Ansley and Kohn's (1990) more recent fast evaluation program. To calculate power, the error covariance matrix, $\Omega(\rho_1, \rho_2)$ in (6.8), or its diagonal component matrix $\Omega_i(\rho_1, \rho_2)$ in (3.27), needs to be decomposed such that

$$\Omega_i(\rho_1, \rho_2) = \Omega_i^{1/2}(\rho_1, \rho_2)\left(\Omega_i^{1/2}(\rho_1, \rho_2)\right)',$$

where the formula for $\Omega_i^{1/2}(\rho_1, \rho_2)$ is given by (3.35).

Observe that the power of the critical region $s < c_\alpha$, in (6.17) can be found by evaluating probabilities of the form,

$$\Pr[s < c_\alpha] = \Pr\left[u'(\Delta - c_\alpha M_X)u < 0 | u \sim N\left(0, \sigma^2 \Omega(\rho_1, \rho_2)\right)\right]$$

$$= \Pr\left[\left(\Omega^{-1/2}(\rho_1, \rho_2)u\right)'\left(\Omega^{1/2}(\rho_1, \rho_2)\right)'(\Delta - c_\alpha M_X)\Omega^{1/2}\right.$$

$$\left. (\rho_1, \rho_2)\left(\Omega^{-1/2}(\rho_1, \rho_2)u\right) < 0 | \Omega^{-1/2}(\rho_1, \rho_2)u \sim N\left(0, \sigma^2 I_n\right)\right]$$

$$= \Pr\left[\sum_{i=1}^{n} \lambda_i \xi_i^2 < 0\right], \tag{6.18}$$

where $\lambda_1, \lambda_2, ..., \lambda_n$ are the eigenvalues of

$$\left(\Omega^{1/2}(\rho_1, \rho_2)\right)'(\Delta - c_\alpha M_X)\left(\Omega^{1/2}(\rho_1, \rho_2)\right), \tag{6.19}$$

and $\xi_1^2, ..., \xi_n^2$ are independent chi-squared random variables each with one degree of freedom. In (6.19), the matrix

$$\Omega^{1/2}(\rho_1, \rho_2) = \bigoplus_{i=1}^{m} \Omega_i^{1/2}(\rho_1, \rho_2),$$

where the elements of the matrix, $\Omega_i^{1/2}(\rho_1,\rho_2)$ is given by the formula (3.35).

Power comparisons of the POI and the LMMPI tests

This section reports a summary of the results of the Wu and Bhatti's (1994) empirical power comparison on the performance of the LMMPI test for intra-block and intra-subblock equicorrelationcoefficients. The POI test discussed in the previous section will be used here in getting the maximum obtainable power at selected points in the alternative parameter space. The differences between this power envelope and the power of the LMMPI test determine the usefulness of the latter test.

Two sets of design matrices are used in this study:

X1 (p=3): Bangladesh agricultural survey data, including a constant, log of labour input per acre in man-days and log of biological-chemical (BC) input per acre in thousand of Bangladesh Taka.[4]

X2 (p=4): Artificially generated data, consisting of a constant, two independent log-normal variables and one uniform random variable. The values of these variables are presented in table 6.3 of appendix 6A.

Here, X1 represents some typical economic survey data whose detail is given in tables 6.1 and 6.2 of appendix 6A. For these data, two block structures are considered. The first structure involves two first-stage blocks with each of them having two subblocks respectively, so $N_i=2$ for i=1,2. The second structure also involves two first-stage blocks, but with each of them having three subblocks respectively, so $N_i=3$ for i=1,2. Also, two sample sizes, n=48, n=96, are considered. So with two block structures and two sample sizes, four cases are associated with X1, X2, respectively:

a. $N_i \times T = 2 \times 12$,   i=1,2,   n =48

b. $N_i \times T = 3 \times 8$,   i=1,2,   n=48

c. $N_i \times T = 2 \times 24$,   i=1,2,   n=96

d. $N_i \times T = 3 \times 16$,   i=1,2,   n=96.

The X1 data follow exactly the above experimental design, whereas the X2 data are properly selected from a set of 96 observations in all cases.

137

The powers are evaluated at points of combinations of ρ_1=0.0, 0.05, 0.1, 0.2, 0.4 and ρ_2=0.0, 0.05, 0.1, 0.2, 0.4. Since ρ_1 and ρ_2 obey the constraint $\rho_1+\rho_2 \leq 1$, and weak to medium equicorrelation is likely, as noted by Scott and Holt (1982) in the two-stage model context, the maximum values of ρ_1 and ρ_2 have been set at 0.4. For each case, the power envelope is evaluated at three points: $(\rho_1, \rho_2)'$=(0.0, 0.2)$'$ (0.2, 0.0)$'$ and (0.2, 0.2)$'$. The critical values and the powers are calculated by using a modified version of Koerts and Abrahamse's (1969) FQUAD subroutine. Table 6.4 presents the results for X1a - X1d, X2a and X2b. Since power patterns in the cases of X2c and X2d are quite similar to those of X1c and X1d, their results are omitted here.

From these tables, it is obvious that the LMMPI test has quite good power for the situations considered. Out of 18 points (three for each case) in the alternative space where the power envelope is calculated, the LMMPI test had power differing from the envelope by less than 0.12 at 17 points, and differing by less than 0.05 at 10 points. Since the LMMPI test has local average optimal power, these calculated powers may be considered reasonable results and are encouraging. Comparing case X1a to X1c and case X1b to case X1d, where sample size has been doubled, power has increased significantly. Also, with this increased sample size, the differences between the power of the LMMPI test and the power envelope have decreased. The amounts are small, but given that these differences are small in the first instance, this decrease is relatively significant. In all cases where the number of subblocks is raised from N_i=2 to N_i=3 while the total number of observations is unchanged, i.e., from case X1a to X1b, X1c to X1d and X2a to X2b, power has dropped for combinations of small to moderately small values of ρ_1 and ρ_2. However, this pattern is reversed for combinations of ρ_1 and ρ_2 with ρ_2=0.4 for cases involving X1 data. For X2 data, this reverse happened at $(\rho_1, \rho_2)'$=(0.4, 0.4)$'$ the furthest point from the null hypothesis considered in our study.

Testing ρ_2 in the presence of ρ_1 [5]

The main purpose of this section is to consider the problem of testing subblock effects in the presence of main block effects by applying King's (1989) approach to the 3SLR model with equicorrelated disturbances within blocks and even within subblocks. Such testing procedures have many applications in physical and social sciences. For example in socioeconomic surveys, a researcher may be interested in testing for the presence of random errors of key punch operators in the presence of coding or an interviewer's error. In time-series cross-sectional data econometricians may be interested in testing firm effects in the presence of industry effects and so on. Note that this testing problem is similar to that of King (1989) who investigated the problem of testing for an AR(4) regression disturbance process in the presence of an AR(1) process. Here the objective is to construct POI, LM1 and LM2 tests to detect the subblock effects in the presence of main block effects.

The model and the tests

The testing procedure is begun by considering the linear regression model (6.6) i.e.,

$$y = X\beta + u, \quad u \sim N\left(0, \sigma^2 \Omega(\rho_1, \rho_2)\right),$$

where

$$\Omega(\rho_1, \rho_2) = \overset{m}{\underset{i=1}{\oplus}} \Omega_i(\rho_1, \rho_2),$$

with submatrices given by (3.27), where $0 \le \rho_1 + \rho_2 \le 1$ and ρ_1, ρ_2 are unknown parameters. For the purpose of test construction, there are a couple of options to decide about the range of ρ_1 values. The first option is that $0 \le \rho_1 \le 1$, which seems unnecessarily generous. Another that might be more reasonable is to assume that $0 \le \rho_1 \le 0.5$, because ρ_1 and ρ_2 follow the constraint $0 \le \rho_1 + \rho_2 \le 1$.

The aim in this section is to test

$$H_0 : \rho_2 = 0, \rho_1 > 0 \tag{6.20}$$

against

$$H_a : \rho_2 > 0, \rho_1 > 0. \tag{6.21}$$

Using invariance arguments, the nuisance parameters except for ρ_1 can be eliminated. Thus the testing problem (6.20) against (6.21) simplifies to one of testing

$$H_0 : \vartheta \text{ has the density } f(\vartheta, \rho_1, 0), \quad 0 < \rho_1 \le 0.5$$

139

against

$$H_a: \vartheta \text{ has the density } f(\vartheta, \rho_1, \rho_2), \quad 0 \leq \rho_1 \leq 0.5, \quad 0 \leq \rho_2 \leq 0.5.$$

Point optimal invariant test

First consider the simple problem of testing the hypothesis

$$H'_0: (\rho_1, \rho_2)' = (\rho_{10}, 0)'$$

against the simple alternative hypothesis

$$H'_a: (\rho_1, \rho_2)' = (\rho_{11}, \rho_{21})',$$

where $0 \leq \rho_{10} \leq 0.5$, $0 \leq \rho_{11} \leq 0.5$ and $0 \leq \rho_{21} \leq 0.5$ are known and fixed.
Suppose under $H_0, \Omega_0 = \Omega(\rho_{10}, 0)$ and under $H_a, \Omega_1 = \Omega(\rho_{11}, \rho_{21})$.

The Neyman-Pearson lemma implies that a Most Powerful (MP) test within the class of invariant tests can be based on critical regions of the form

$$s(\rho_{10}, \rho_{11}, \rho_{21}) = \frac{\vartheta'(P\Omega_1 P')^{-1}\vartheta}{\vartheta'(P\Omega_0 P')^{-1}\vartheta} < c_\alpha$$

where c_α is an appropriate critical value. King (1989, p. 290) showed that the statistics $s(\rho_{10}, \rho_{11}, \rho_{21})$ can also be written as

$$s(\rho_{10}, \rho_{11}, \rho_{21}) = \frac{u'\Delta_1 u}{u'\Delta_0 u} \tag{6.22}$$

where

$$\Delta_1 = \Omega_1^{-1} - \Omega_1^{-1} X (X'\Omega_1^{-1} X)^{-1} X'\Omega_1^{-1}$$

$$\Delta_0 = \Omega_0^{-1} - \Omega_0^{-1} X (X'\Omega_0^{-1} X)^{-1} X'\Omega_0^{-1}.$$

Note that $s(\rho_{10}, \rho_{11}, \rho_{21})$ in (6.22) is a ratio of quadratic forms in normal variables, therefore, its critical values, c_α, for a desired level of significance, α, may be obtained by evaluating the probabilities of the form

$$\Pr\left[s(\rho_{10}, \rho_{11}, \rho_{21}) < c_\alpha \middle| u \sim N(0, \Omega_0)\right]$$

$$= \Pr\left[u'(\Delta_1 - c_\alpha \Delta_0)u < 0 \middle| u \sim N(0, \Omega_0)\right]$$

$$= \Pr\left[\left(\Omega_0^{-1/2}u\right)'\left(\Omega_0^{1/2}\right)'(\Delta_1 - c_\alpha \Delta_0)\Omega_0^{1/2}\left(\Omega_0^{1/2}u\right) < 0 \middle| u \sim N(0, \Delta_0)\right]$$

$$= \Pr\left[\sum_{i=1}^{n} \lambda_i \xi_i^2 < 0\right], \qquad (6.23)$$

where $\lambda_1, \ldots, \lambda_n$ are eigenvalues of the matrix

$$\left(\Omega_0^{1/2}\right)'(\Delta_1 - c_\alpha \Delta_0)\left(\Delta_0^{1/2}\right),$$

and $\xi_1^2, \ldots, \xi_n^2$ are independent chi-squared random variables each with one degree of freedom.

Thus a test based on $s(\rho_0, \rho_1, \rho_2)$ in (6.22) is most powerful invariant in the neighbourhood of $(\rho_1, \rho_2)' = (\rho_{11}, \rho_{21})'$. Therefore it is a POI test of the simple null H_0' against the simple alternative H_a'. For the wider problem of testing (6.20) against (6.21) it may not necessarily by POI, because for this testing problem the critical value should be found by solving

$$\sup_{0 < \rho_1 \le 0.5} \Pr\left[s(\rho_{10}, \rho_{11}, \rho_{21}) < c_\alpha^* \middle| u \sim N(0, \Omega(\rho_1, 0))\right] = \alpha \quad \text{for } c_\alpha^*.$$

In general, $c_\alpha < c_\alpha^*$ so the two critical regions are different. As the range of ρ_1 is closed

$$\Pr\left[s(\rho_{10}, \rho_{11}, \rho_{21}) < c_\alpha^* \middle| u \sim N(0, \Omega(\rho_1, 0))\right] = \alpha$$

141

is true for at least one ρ_1 value. If we can find such a ρ_{10} value to let $c_\alpha = c_\alpha^*$ then the test based on $s(\rho_{10}, \rho_{11}, \rho_{21})$ is MPI and is a POI test of (6.20) against (6.21).

However, in general, there is no reason to believe that such a ρ_{10} value should exist. Perhaps it may exist for some combinations of X matrices and $(\rho_{11}, \rho_{21})'$ values, but not for others. For the X matrices used in this experiment such a ρ_{10} does exist and hence the POI test also exists for testing (6.20) against (6.21).

Once the value of $\rho_{10} = \rho_{10}^*$ (say) is chosen, the optimal values of ρ_{11}, ρ_{21} can be decided upon in the same way as in King (1989). This implies that for any fixed ρ_{21}, the power of the test will be determined in part by the choice of ρ_{11}. We suggest choosing ρ_{11} to maximize the minimum power in the alternative parameter space i.e.

$$\max_{0 \le \rho_1 \le 0.5} \left[\min \Pr \left\{ s(\rho_{10}^*, \rho_{11}, \rho_{21}) < c_\alpha^* | u \sim N(0, \Omega(\rho_{11}, \rho_{21})) \right\} \right].$$

This ρ_{11} value is denoted by ρ_{11}^*. Then the optimal value of ρ_{21} can be found by solving

$$\Pr \left[s(\rho_{10}^*, \rho_{11}^*, \rho_{21}) < c^* | u \sim N(0, \Omega(\rho_{11}^*, \rho_{21})) \right] = p, \tag{6.24}$$

for $\rho_{21} = \rho_{21}^*$, where p on the right hand side of (6.24) denotes the requisite level of the power. It may be noted that a simpler form of this exercise in the context of a BO test has already been done in chapter three. This is another example of applying the same logic to a more complex set of given parameters under the alternative parameter space.

In the next section the procedure of calculating critical values and the power of the POI test for some selected design matrices will be discussed. This is done in order to compare the power performance of a POI test with its other competitive tests, namely LM1 and LM2 tests, as it has been done in chapter five. In the next subsection the derivation of these tests will be demonstrated and the power comparison will be shown in a later section of this chapter.

The Lagrange multiplier test is generally a two-sided test and is only valid asymptotically. In other words it relaxes the constraints on the sign of the parameters under the alternative hypothesis. The main advantage of the LM test over its asymptotic equivalent tests, namely, the LR and Wald tests is that it require an estimate of the value of the parameters only under the null hypothesis. Due to this it has gained wide acceptance from practitioners in the econometrics literature. Recent examples include Breusch and Pagan (1980), Honda (1985), Godfrey (1988), Baltagi, Chang and Li (1991) and Wu (1991). For a recent survey of general econometric methods, see Pagan and Wickens (1989). Krämer and Sonnberger (1986) provide some detailed descriptions of various tests in the context of the linear regression model. Here in this section the LM1 and LM2 tests will be considered because as was noted they just require the estimation of ρ_1 under the null hypothesis.

Corresponding to the testing problem of (6.20) against (6.21), consider the problem of testing

$$H_0':\rho_2 = 0, \quad \rho_1 \neq 0$$

against

$$H_a':\rho_2 \neq 0, \quad \rho_1 \neq 0.$$

The log likelihood function for the model (6.6) is

$$L(\beta,\sigma^2,\rho_1,\rho_2) = -\frac{n}{2}\ell n(2\pi\sigma^2) - \frac{1}{2}\ell n|\Omega(\rho_1,\rho_2)|$$
$$- \frac{1}{2\sigma^2}(y - X\beta)'\Omega^{-1}(\rho_1,\rho_2)(y - X\beta). \tag{6.25}$$

It can be verified that

$$\frac{\partial L}{\partial \rho_2}\bigg|_{H_0} = d_1 - \frac{n}{2}\frac{\hat{u}'A\hat{u}}{\hat{u}'\hat{u}} \tag{6.26}$$

in which $d_1 = -\frac{1}{2}\frac{\partial|\Omega(\rho_1,\rho_2)|}{\partial \rho_2} \cdot |\Omega(\rho_1,\rho_2)|^{-1}\big|_{H_0}$

or equivalently,

$$\frac{n(T-1)b}{2},$$

where

$$b = \frac{-\rho}{(1-\rho_1)(1-\rho_1+sT\rho_1)},$$

and A in (6.26) is expressed as,

$$A = \Omega^{-1/2}(\rho_1,0)(I_n - D_2)\left(\Omega^{-1/2}(\rho_1,0)\right)'.$$

The information matrix ψ_θ of the linear regression model (6.6) can be obtained by using Magnus (1978, theorem 3), where ψ_θ is a symmetric 4x4 matrix with typical element

$$(\psi_\theta)_{ij} = \frac{1}{2}\mathrm{tr}\left(\frac{\partial \Omega^{-1}}{\partial \theta_i}\Omega\frac{\partial \Omega^{-1}}{\partial \theta_j}\Omega\right).$$

Let $\Omega(\theta) = \sigma^2\Omega(\rho_1,\rho_2)$, where $\theta = (\theta_1,\theta_2,\theta_3)' = (\sigma^2,\rho_1,\rho_2)'$.

Thus the inverse of our block diagonal information matrix evaluated at the null, is

$$\psi^{-1}\big|_{H_0} = \begin{bmatrix} (x'\Omega^{-1}x)^{-1} & 0 & 0 & 0 \\[2mm] \begin{matrix} 0 \\ 0 \\ 0 \end{matrix} & \begin{bmatrix} \psi_{11} & \psi_{12} & \psi_{13} \\ \psi_{21} & \psi_{22} & \psi_{23} \\ \psi_{31} & \psi_{32} & \psi_{33} \end{bmatrix}^{-1} \end{bmatrix}_{H_0}.$$

Let I_0 be denoted as the bottom right element of ψ^{-1} which corresponds to ρ_2, i.e.

$$I_0 = \left[\psi_{33} - \begin{pmatrix}\psi_{31} & \psi_{32}\end{pmatrix}\begin{bmatrix}\psi_{11} & \psi_{12} \\ \psi_{12} & \psi_{22}\end{bmatrix}^{-1}\begin{pmatrix}\psi_{13} \\ \psi_{23}\end{pmatrix}\right]^{-1}_{H_0},$$

which can be simplified as

$$I_0 = \left[\psi_{33} - \frac{1}{\left(\psi_{11}\psi_{22} - \psi_{12}^2\right)} \left\{ \psi_{13}^2 \psi_{22} - 2\psi_{12}\psi_{13}\psi_{23} + \psi_{11}\psi_{23}^2 \right\} \right]^{-1} \Bigg|_{H_0} \tag{6.27}$$

Hence the I_0 in (6.27) is obtained by noting that

$$\psi_{ij} = \psi_{ji}, \quad i,j = 1,2,3, \quad \text{for} \quad i \neq j, \quad i < j.$$

Also note that the elements of ψ_{ij} under the null hypothesis are defined as follows:

$$\psi_{11} = \frac{n}{2\sigma^4}, \quad \psi_{12} = \frac{nb(sT-1)}{2\sigma^2}, \quad \psi_{13} = \frac{nb(T-1)}{2\sigma^2}$$

$$\psi_{22} = \frac{n(1-sT)}{2}\left[-a^2 + b^2 sT(1-sT) + 2ab(1-sT)\right],$$

$$\psi_{32} = \frac{n(1-T)}{2}\left[-a^2 + b^2 sT(1-sT) + 2ab(1-sT)\right],$$

$$\psi_{33} = \frac{n(1-T)}{2}\left[-a^2 + b^2 sT(1-T) + 2ab(1-T)\right], \tag{6.28}$$

where

$$a = \frac{1}{1-\rho_1} \quad \text{and b is defined previously by (6.26).}$$

Thus, by (6.26), (6.27) and (6.28), the LM1 test statistic is

$$s1 = \frac{\partial L}{\partial \rho_2}\Big|_{H_0} / \sqrt{I_0}, \tag{6.29}$$

and the LM2 test statistic is

$$s2 = \left(\frac{\partial L}{\partial \rho_2}\Big|_{H_0}\right)^2 \Big/ I_0. \tag{6.30}$$

145

The asymptotic distributions of the s1 and s2 test statistics under H_0 are standard normal and chi-square with one degree of freedom, respectively. Note that under H_0, the ρ_1, σ^2 and β are replaced by their respective maximum likelihood estimates, namely $\hat{\rho}_1$, $\hat{\sigma}^2$ and $\hat{\beta}$. Therefore, the ψ_{ij}'s will also be replaced by $\hat{\psi}_{ij}$'s for all values of i and j defined above. The next section will report the results of the Monte Carlo study designed to compare the empirical power of the one-sided and two-sided Lagrange multiplier tests with that of the two versions of the POI test for testing intra-subblock equicorrelation in the presence of intrablock equicorrelation for the various selected design matrices.

Powers and sizes comparison of LM1, LM2 and POI tests

This section reports the results of Monte Carlo experiments which were conducted to assess and compare the small-sample size and power performance of the two versions of the POI tests, namely $s(\rho_{10}^*, 0.1, 0.1)$, $s(\rho_{10}^*, 0.25, 0.25)$ and the LM1 and the LM2 tests of testing null hypothesis of (6.20) against an alternative (6.21). The main objectives of this analysis were to assess the validity (or accuracy) of the asymptotic critical values for each test in finite samples and to compare the power curves of the asymptotic tests (i.e. LM1 and LM2) with the two versions of the POI tests.

Experimental design and the computations

This experiment is divided into two parts. The first part used the Monte Carlo method to estimate probabilities of a type I error under H_0 at $\rho = 0.0$, 0.05, 0.1, 0.2, 0.3, 0.4 and 0.5, for the LM1 and LM2 tests at the five percent nominal level. The second part of the experiment is completed in two stages. The first stage being the calculation of appropriate critical values so that the comparison of powers, which constitute the second stage, can be made at approximately the same level of significance. Critical values of ρ^*_{10} for the POI tests were calculated according to the procedure given previously for all X matrices defined below. Their calculated values are tabulated in table 6.5 of appendix 6B. For the LM1 and LM2 tests, the Monte Carlo method was used to estimate exact-size critical values at $\rho_1 = $ 0.0, 0.05, 0.1, 0.2, 0.3, 0.4 and 0.5 under H_0 at the five percent level. From each set of seven critical values, the largest was selected, thus ensuring that, at least at the chosen points, the size of the test does not exceed five percent.

The small-sample sizes and powers were then calculated using the methodology discussed previously for the POI tests, whereas the methodology for the LM1 and LM2 tests is given earlier in the subsection entitled Lagrange Multiplier Test. Sizes were calculated for the equicorrelated parameter, ρ_1, values 0.0, 0.05,...,0.5 and powers were calculated at $\rho_1=\rho_2=0.0$, 0.05,...,0.5. Finally, for the small-samples, the sizes and the powers of the $s\left(\rho_{10}^*,0.1,0.1\right)$ and $s\left(\rho_{10}^*,0.25,0.25\right)$ tests were calculated at ρ_1 and ρ_2 values of 0.0, 0.05, 0.1, 0.2, 0.3, 0.4 and 0.5. In the remainder of this section these tests will be denoted as the s(0.1, 0.1) and s(0.25, 0.25) tests.

The following design matrices were used in this study:

Bangladesh agricultural survey data,[6]

X1: (nx3; n=24, m=2, s=3, T=4)

     (nx3, n=48, m=2, s=3, T=8)

Australian data[7]

X2: (nx3; n=24, m=2, s=3, T=4)

     (nx3; n=64, m=2, s=4, T=8)

Artificially generated data

X3: (nx3; n=24, m=2, s=3, T=4)

     (nx3; n=72, m=2, s=3, T=12).

These design matrices reflect a variety of economic phenomenon and have also been used in earlier empirical studies (see Hoque:1988, 1991; King and Evans: 1986; Wu and Bhatti: 1994; and Wu: 1991 among others).

The Monte Carlo method was applied using two thousand replications for the lm tests. The results of Breusch (1980) imply that the sizes and powers of each of the tests are invariant to the values taken by β and σ^2. Therefore, for the application of the Monte Carlo method, β_i, i=1,2,...,p and σ^2 were all set to unity. Whenever needed, the innovations, v_t, were generated as pseudo-random normal variables as described by King and Giles (1984). Then these innovations were transformed to the equicorrelated error

147

structure of the form u_{ijk}, defined by (6.1) using a subroutine under H_0, following the logic of Beach and MacKinnon's (1978) but otherwise different from them.[8] The maximum likelihood estimates of β, σ^2, and ρ_1 under H_0 were computed using Ansley's (1979) approach for estimating ARMA process applied to the 2SLR model. For an analogous application of Ansley's algorithm to the linear regression model and the 2SLR model with varying coefficients, see King (1986) and Bhatti (1993), respectively. This reduces the maximization problem to a sum of squares minimization problem which was handled by the IMSL subroutine DBCLSF from the IMSL MATH/LIBRARY (1989), with the constraint that $|\rho_1| \le 1$. For the POI test, the probabilities of the form (6.23) and (6.24) were calculated using the eigenvalue method and a modified version of Koerts and Abrahamse's (1969) FQUAD subroutine with maximum integration and truncation errors of 10^{-6}.

The results

Tables 6.6 and 6.7 in appendix 6B report the estimated sizes for the various sample-sizes over ρ_1=0.0, 0.05, 0.1, 0.2, 0.3, 0.4, 0.5 of the LM1 and LM2 test statistics using asymptotic critical values at the five percent significance level. In these tables, it can be noticed that almost all of the estimated sizes of the LM1 and LM2 tests for the different sample sizes are significantly different than the nominal sizes of 0.05 for the X1, X2 and X3 design matrices. The exceptions are only in table 6.6, for the design matrices X1 for ρ_1=0.05, 0.1, 0.2, X2 for ρ_1=0.0, 0.5 and X3 for ρ_1=0.0, 0.05 and 0.1, where the estimated sizes of the LM2 test do not differ significantly from the nominal size. In all cases, the estimated sizes of the LM2 test are more than half of the nominal level, except in table 6.7 where for the design matrix X2 for ρ_1=0.4, 0.5 and X3 for ρ_1=0.5 it is less by 0.002, 0.003 and 0.001, respectively. The estimated sizes of the LM1 test for the 100 per cent of the cases in table 6.6 and 6.7 differ significantly than the five per cent nominal size, for the sample sizes considered in this experiment.

The sizes of the LM1 test have the tendency to increase with the increase in the sample size, whereas the sizes of the LM2 test decreases. Generally, the sizes remain less than the five percent nominal level. The lowest sizes of the LM1 test for X1, n=24 and n=48 are 0.013 and 0.019 whereas for X2, n=24 and n=64 they are 0.010 and 0.017, respectively. While the lowest estimated sizes for X3, n=24 and n=72 are 0.018 and 0.023, respectively. The highest estimated sizes of the LM1 test for X1, n=24 and n=48 are 0.018 and 0.023, for X2, n=24 and n=64 are 0.013 and 0.019 whereas for X3, n=24 and n=72 they are 0.023 and 0.028, respectively. Thus the sizes of the LM1 test increases as the sample sizes increases. While the sizes of the LM2 test behave opposite to that of the sizes of the LM1 test.

The rest of this subsection will concentrate on discussing the power behaviour of the two versions of the POI tests namely, s(0.1, 0.1) and s(0.25, 0.25), the LM1 and the LM2 tests. The power calculations of all these tests for the various design matrices, different sample-sizes and different block-subblock-sizes are presented in tables 6.8 to 6.13.

In general, under H_a, the power function of all the tests increases with the increase in the sample size, subblock size, the number of subblocks in the main block and the values of ρ_1 and ρ_2. The power of the POI tests, for small-sample sizes (n=24) are always greater than or equal to the power of the LM1 test. Exceptions occur in table 2.9 for $\rho_1=0.1, 0.2$ and $\rho_2=0.05$ where the power of LM1 test is a bit larger than that of POI tests by 0.002, which is of course not significant. It is also noticeable in table 6.11, as the sample size increases, the power of the LM1 test is becoming closer to the POI tests for higher values of ρ_1 and ρ_2. While for the extremely lower values of ρ_1, the power of the LM1 test is marginally better than its rival POI test, but only just. However, the LM2 test has always the lowest power among all the tests under consideration.

Generally speaking, the power difference between POI tests and the LM1 test is significant for all values of $\rho_1, \rho_2>0.2$. The exceptions in table 6.8 are for $\rho_1=\rho_2=0.1$, $\rho_1=0.2$, $\rho_2=0.1$ and $\rho_1=0.0, 0.05, 0.1, 0.2$, with $\rho_2=0.2$ and in table 6.12 for $\rho_1=\rho_2=0.2$. For the rest of the combination of the values of ρ_1, ρ_2, which are ≤ 0.2, the power of the LM1 and POI tests is approximately similar.

Finally, we turn to the comparison of powers of the two versions of the POI tests, namely the s(0.1, 0.1) and s(0.25, 0.25) tests of H_0 against H_a. All calculated powers of these tests for X1, X2 and X3 design matrices are prescribed in the first two rows of the tables 6.8 to 6.13 of appendix 6B. These tables reveal that for the small-samples (i.e., n=24) in tables 6.8 to 6.10 the maximum power difference between these tests ranges from 0.001 to 0.035. For this case it seems that the s(0.25, 0.25) test have the best overall power properties. For the moderate sample sizes in tables 6.11, 6.12 and 6.13, the powers of the s(0.1, 0.1) and s(0.25, 0.25) tests are approximately similar. The maximum power difference between both tests is 0.003. Therefore, this section concludes that for moderate sample sizes the POI test is approximately UMPI along $\rho_1=\rho_2$, because its power curve is approximately unchanged for the different values of the parameters under the alternative parameter space.

149

Multi-stage model and the tests

The model

This section extends the three-stage model considered in an earlier section to the situation where observations are obtained from a multi-stage block design. In this case, the ith block (say, states) is divided into subblocks (districts), sub-subblocks (suburbs), sub-sub-subblocks (major street blocks) and so on. Thus, it obtains a total of n observations which are sampled from m first-stage states, with m(i) districts from the ith state, with m(i,j) suburbs from the jth district of the ith state, with m(i,j,k) major street blocks from the kth suburbs of the jth district of the ith state and so on, with m(i,j,k,l,...,q,r) observations from the qth household of the ..., lth street block, of the kth suburb of the jth district of the ith state, such that the sample size

$$n = \sum_{i=1}^{m} \sum_{j=1}^{m(i)} \sum_{k=1}^{m(i,j)} \cdots \sum_{r=1}^{m(i,j,...,p')} m(i,j,k,...,p',q).$$

Thus, the multistage linear regression model at the τth level can be expressed as

$$y_{ijk...p'qr} = \sum_{\ell=1}^{p} \beta_{\ell} x_{(ijk...p'qr)\ell} + u_{ijk...p'qr} \tag{6.31}$$

for observations r=1,2,...,m(i,j,k,...,q) from the qth household, households q=1,2,...,m(i,j,...,p'),...., from the kth suburb, suburbs k=1,2,...,m(i,j), from the jth district, districts j=1,2,...,m(i), from the ith state for states i=1,...,m with $y_{ijk...r}$ as the dependent variable and p independent variables $x_{(ijk...p'q)\ell}$ ℓ=1,...,p, one of which may be a constant. The error term $u_{ijk...r}$ can be written as

$$u_{ijk...qr} = v_i + v_{ij} + v_{ijk} + ... + v_{ijk...qr},$$

where v_i is the state (block) effects, v_{ij} is the jth district effect in the ith state, v_{ijk} is the kth suburb effect in the jth district of the ith state and so on. The $v_{ijk...qr}$ is the remaining random effect. These r components of $u_{ijk...qr}$ are assumed to be mutually independent and normally distributed with

$$E(v_i) = E(v_{ij}) = ... = E(v_{ijk...qr}) = 0 \tag{6.32}$$

150

and

$$\text{var}(v_i) = \sigma_1^2, \ \text{var}(v_{ij}) = \sigma_2^2, \ldots, \ \text{var}(v_{ijk\ldots qr}) = \sigma_\tau^2,$$

so that $E(u_{ijk\ldots qr}) = 0$ and

$$E(u_{ijk\ldots qr} \ u_{i'j'k'\ldots q'r'}) = \begin{cases} 0 & \text{for } i \neq i' \text{ any } j,j',k,k',\ldots,q,q',r,r' \\ \sigma_1^2 & \text{for } i = i' \ j \neq j', \text{ any } k,k',\ell,\ell',\ldots q'r,r' \\ \sigma_1^2 + \sigma_2^2 & \text{for } i = i', j = j', k \neq k', \text{ any } \ell,\ell\ldots q,q',r,r' \\ \sigma_1^2 + \sigma_2^2 + \sigma_3^2 & \text{for } i = i', j = j', k = k', \ell = \ell, \ldots \text{any } q,q',r,r' \\ \vdots \\ \sigma_1^2 + \sigma_2^2 + \sigma_3^2 + \ldots + \sigma_{\tau-1}^2 & \text{for } (i,j,k,\ldots,q) = (i',j',k',\ldots,q'), r \neq r' \\ \sigma_1^2 + \sigma_2^2 + \ldots + \sigma_{\tau-1}^2 + \sigma_\tau^2 & \text{for } (i,j,k,\ldots,q) = (i',j',\ldots,q',r') \end{cases}$$

(6.33)

for   r,r´=1,2,...,m(i,j,k,...,q)
     q,q´=1,2,...,m(i,j,k,...,p')

k,k´=1,2,...,m(i,j)
j,j´=1,2,...,m(i)
i,i´=1,2,....,m.

If $\sigma_1^2 + \sigma_2^2 + \ldots + \sigma_\tau^2 = \sigma^2$, then the coefficient of intra-state equicorrelation is defined as

$$\rho_1 = \sigma_1^2 / \sigma^2.$$

The equicorrelation coefficient of intra-district and the intra-suburb are

$$\rho_2 = \frac{\sigma_1^2 + \sigma_2^2}{\sigma^2} \quad \text{and} \quad \rho_3 = \frac{\sigma_1^2 + \sigma_2^2 + \sigma_3^2}{\sigma^2},$$

respectively, and so on.

Generally, the j'th intra-subdivision equicorrelation is defined as

$$\rho'_j = \sum_{i=1}^{j'} \sigma_i^2 / \sigma^2, \quad \text{for } j' = 1,...,(\tau-1).$$ (6.34)

The regression model (6.31) under (6.32), (6.33) and (6.34) can be written in matrix form as

$$y = X\beta + u$$ (6.35)

in which y and u are $n \times 1$, β is $p \times 1$ and

$$u \sim N\left(0, \sigma^2 \Omega(\rho_1, \rho_2, ..., \rho_{\tau-1})\right),$$

where

$$\Omega(\rho_1, \rho_2, ..., \rho_{\tau-1}) = \bigoplus_{i=1}^{m} \Omega_i(\rho_1, \rho_2, ..., \rho_{\tau-1}),$$ (6.36)

block diagonal with submatrices $\Omega_i(\bullet)$. For simplicity, only the case of balanced data will be considered. If $s_1 = m(i)$, $s_2 = m(i,j)$,, $s_3 = m(i,j,k)$,..., $s_{p'} = m(i,j,k,...,p')$ $s_q = m(i,j,k,...,l)$. Then (6.36) can be simplified as,

$$\Omega(\rho_1, ..., \rho_{\tau-1}) = \left(1 - \sum_{i=1}^{\tau-1} \rho_i\right) I_n + \sum_{i=1}^{\tau-1} \rho_i D_i$$ (6.37)

where

$$D_1 = I_m \otimes E_{(s_1 s_2 .... s_q)}$$

$$D_2 = I_{m s_1} \otimes E_{(s_2 s_3 .... s_q)}$$

$$D_3 = I_{m s_1 s_2} \otimes E_{(s_3 s_4 .... s_q)}$$

$$\vdots \qquad \vdots$$ (6.38)

$$D_q = I_{(m s_1 s_2 .... s_q)} \otimes E_{s_q}.$$

The testing problem and the tests

This subsection is interested in testing

$$H_0: \rho_i = 0, \quad \text{against} \tag{6.39}$$

$$H_a: \rho_i > 0, \quad \text{for } i = 1, \dots, \tau - 1, \text{ (excluding } H_0). \tag{6.40}$$

Note that this testing problem is a general form of testing (6.9) against (6.10). Therefore, this problem is also invariant under the general group of transformations of the form (6.11), with maximal invariant vector ϑ, given by (6.12). The probability density function of ϑ in this case, under (6.36) and (6.37) can be written in the form of (6.13) as follows.

$$f\left(\vartheta, \rho_1, \dots, \rho_{\tau-1}\right) = 1/2\Gamma\left(\frac{g}{2}\right)\pi^{-g/2}\left|P\Omega(\rho_1, \dots, \rho_{\tau-1})P'\right|^{-1/2}$$

$$\left(\vartheta'\left(P\Omega(\rho_1, \dots, \rho_{\tau-1})P'\right)\vartheta\right)^{-g/2}d\vartheta, \tag{6.41}$$

where $d\vartheta$ denotes the uniform measure on the unit sphere and $g=n-p$, $M_x=I_n-X(X'X)^{-1}X'$ and P is an gxn matrix such that $PP'=I_g$ and $P'P=M_x$. Note that $\rho_1, \rho_2, \dots, \rho_{\tau-1}$ are the unknown parameters in (6.41).

Point optimal invariant test

This subsection chooses $(\rho_1, \rho_2, \dots, \rho_{(\tau-1)})' = (\rho_{11}, \dots, \rho_{21}, \rho_{(\tau-1)}, 1)'$ a point in the alternative parameter space. Based on the maximal invariant (6.12), with the density (6.41), one can use King's (1987b) equation (18) in constructing the POI test for testing (6.39) against (6.40), that is to reject H_0 for small values of

$$\bar{s} = \frac{\hat{u}'\Delta\hat{u}}{\hat{u}'\hat{u}} \tag{6.42}$$

where

$$\Delta = \Omega_1^{-1} - \Omega_1^{-1}X\left(X'\Omega_1^{-1}X\right)^{-1}X'\Omega_1^{-1}$$

153

$$\Omega_1 = \Omega\left(\rho_{11}, \rho_{21}, \ldots, \rho_{(\tau-1)1}\right)$$

and $\hat{u}$ is the OLS residual vector, $\hat{u} = M_x y$. Thus a test based on $\bar{\bar{s}}$ gives the maximum attainable power at $\left(\rho_{11}, \rho_{21}, \ldots, \rho_{(\tau-1)1}\right)$ in the class of invariant tests. Further note that, like s in (6.17), $\bar{\bar{s}}$ in (6.42) is also a ratio of quadratic forms in multi-normal variates. Therefore, its critical values and power function can be solved by using the standard numerical algorithm mentioned in earlier chapters.

Locally most mean powerful invariant test

As was noticed in the previous subsection that the general form of testing (6.39) against (6.40) is invariant under the group of transformations of the form (6.11), with the maximal invariant vector, ϑ, of the form (6.12) and the form of the density of ϑ could be obtained from (6.41). Based on the maximal invariant vector of the form (6.12), when more than one parameter are being tested, one can apply King and Wu's (1990) theorem to develop a LMMPI test for testing (6.39) against (6.40). That is to reject H_0, i.e. (6.39), for small values of

$$\bar{\bar{s}} = \frac{\hat{u}'A\hat{u}}{\hat{u}'\hat{u}},$$

where

$$A = (\tau-1)I_n - \sum_{i=1}^{\tau-1} D_i,$$

which is a generalized form of A, in (6.15). Here D_i for $i=1,2,\ldots,(\tau-1)$, is already been defined by (6.38). If one wishes one can also write (6.43) in a simplified form like that of (6.16). Further, note that the critical values and the power of this test can also be calculated by using the standard numerical techniques mentioned earlier.

Concluding remarks

In this chapter the problem of testing for equicorrelation in linear regression errors due to multi-block effects was developed. For an example, the three-stage linear regression model was considered and a POI and LMMPI test for testing block/subblock effects for this model was constructed. The power study reported in the section entitled 'Power comparison of the POI and the LMMPI tests' showed that the LMMPI test had good power for the data sets used, and so gave support for its usefulness in developing these tests for testing multi-block effect in the previous section. The second issue to be considered in this chapter was that the testing for subblock effects in the presence of main block effects. An exact POI test and the asymptotic LM tests for this more complicated problem was developed. The exact sizes and powers of the two versions of the POI test were computed using the method illustrated in the section entitled 'Testing ρ_2 in the presence of ρ_1' while those of the asymptotic tests, namely LM1 and LM2 were estimated using the Monte Carlo simulation method.

The results obtained from detailed calculations reported in appendix 6B revealed that in most of the cases, estimated sizes of the LM1 test are less than or equal to the nominal size of 0.05. The sizes of the LM1 test have the tendency to increase as the sample size increases. It was found that the POI test is marginally better than the LM1 test for small and moderate sample sizes. It was also noticed that the POI test is approximately UMPI. Finally, based on these encouraging power results in this study the POI and the LMMPI tests for testing multi-block effects while dealing with the multi-stage linear regression model have been generalised. Furthermore, if one wishes, one can extend the testing problem of the section entitled 'Testing ρ_2 in the presence of ρ_1' to a more general form of testing lower level block effects in the presence of higher level block effects.

Notes

1   Some of the findings reported in the section entitled 'Three-stage linear regression model and tests' of this chapter were published in Wu and Bhatti (1994). The derivation of the LMMPI test for block effects were done jointly by Dr. Ping Wu and the author. The related empirical power calculations in the section entitled 'Point optimal invariant test' were planned jointly and carried out by Dr. Ping Wu.

2   u is formed as u= $(u_{111}, u_{112}, \ldots u_{mm(i)m(i,j)})'$ where X and y can also be formed in a similar pattern.

3   That is, different number of observations in each subblock and different number of subblocks in each main block.

4   Thanks are due to Dr. Asraul Hoque for making this data available.

5   Some of the findings reported in this section were presented at the 1992 Australasian Meeting of the Econometrics Society, Department of Econometrics, Monash University, Clayton, Australia, July 6-8, 1992. The title of the paper was 'Testing for subblock effect in the multistage linear regression models'.

6   The values of the variables, details of the sampling design and measurement of variables are given in appendix 6A. This design matrix has also been used in the empirical study of the section entitled 'Power comparison of the POI and LMMPI tests' with sample of size 96.

7   The detailed explanation of this design matrix is already given in chapter five.

8   Their subroutine deals with autocorrelated errors, mines is concerned with equicorrelated disturbances for the 3SLR model.

Appendix 6A: The data used in this chapter

The symbols used here are described on page 137 of this chapter, where i=1,2,...,m, refer to the number of blocks, j=1,2,...,N$_j$=s, refer to the number of subblocks in the ith block and k=1,2,...,T, the number of observations in the jth subblock of the ith block.

Table 6.1
X1: Bangladesh agricultural data for n=96, m=2, s=3, T=16

i	j	k	x_{ijk2}	x_{ijk3}	i	j	k	x_{ijk2}	x_{ijk3}
1	1	1	2.0	13.0	1	2	9	3.8	14.7
1	1	2	5.0	14.0	1	2	10	4.9	14.6
1	1	3	1.9	13.0	1	2	11	7.0	13.0
1	1	4	2.3	12.8	1	2	12	8.5	12.1
1	1	5	7.0	13.5	1	2	13	6.6	13.8
1	1	6	6.5	14.0	1	2	14	8.0	12.6
1	1	7	4.0	14.5	1	2	15	11.2	12.0
1	1	8	3.8	14.6	1	2	16	9.3	11.9
1	1	9	4.5	14.4					
1	1	10	5.0	14.8	1	3	1	2.0	13.2
1	1	11	9.0	12.2	1	3	2	5.0	14.5
1	1	12	11.0	12.0	1	3	3	3.8	14.0
1	1	13	15.0	11.0	1	3	4	4.0	14.3
1	1	14	6.0	14.0	1	3	5	5.5	14.2
1	1	15	9.0	12.0	1	3	6	4.7	14.6
1	1	16	12.0	11.5	1	3	7	6.8	13.8
					1	3	8	2.5	13.6
1	2	1	1.9	13.0	1	3	9	3.9	14.0
1	2	2	5.0	14.2	1	3	10	5.0	14.6
1	2	3	3.7	14.6	1	3	11	6.9	13.8
1	2	4	4.0	14.8	1	3	12	8.4	12.9
1	2	5	21.6	10.3	1	3	13	6.6	13.9
1	2	6	4.6	14.3	1	3	14	8.0	13.0
1	2	7	7.8	12.7	1	3	15	10.5	12.6
1	2	8	2.5	13.9	1	3	16	9.4	12.8

157

Table 6.1 continued

i	j	k	x_{ijk2}	x_{ijk3}	i	j	k	x_{ijk2}	x_{ijk3}
2	1	1	4.5	14.8	2	2	9	2.8	13.7
2	1	2	3.9	14.4	2	2	10	4.0	14.2
2	1	3	5.0	15.0	2	2	11	6.5	14.5
2	1	4	6.0	15.1	2	2	12	8.0	14.1
2	1	5	3.7	14.4	2	2	13	7.0	14.0
2	1	6	4.0	14.5	2	2	14	9.2	13.8
2	1	7	8.0	14.7	2	2	15	5.5	14.4
2	1	8	9.5	14.3	2	2	16	10.0	13.9
2	1	9	11.0	14.2					
2	1	10	10.0	14.0	2	3	1	1.9	13.9
2	1	11	6.5	15.0	2	3	2	3.0	14.2
2	1	12	2.9	14.2	2	3	3	2.5	14.0
2	1	13	7.7	14.8	2	3	4	4.0	14.4
2	1	14	12.0	13.9	2	3	5	5.2	15.0
2	1	15	8.5	14.7	2	3	6	7.0	14.8
2	1	16	1.9	14.0	2	3	7	4.6	14.9
					2	3	8	6.0	15.2
2	2	1	2.5	13.7	2	3	9	5.5	15.2
2	2	2	4.0	14.1	2	3	10	7.0	15.0
2	2	3	1.7	13.4	2	3	11	2.9	14.2
2	2	4	3.7	14.0	2	3	12	5.0	15.0
2	2	5	5.0	14.5	2	3	13	8.0	14.5
2	2	6	4.3	14.2	2	3	14	6.4	15.1
2	2	7	1.9	13.5	2	3	15	4.8	14.9
2	2	8	5.3	14.3	2	3	16	11.0	14.2

The data selected in table 6.1 is based on multistage block sampling design for the year 1986-87. The primary sampling units or first-stage sampling units are the two main administrative regions of Bangladesh, namely Khulna and Rajshahi divisions, from which four and three districts (subblocks) are randomly selected, respectively. From each district a number of farms, proportional to the cultivated area of the district are selected. There has been 600 farms collected from Khulna and Rajshahi divisions. The districtwise farm distribution of each division is given in table 6.2:

158

Table 6.2
The sampling design and the districtwise farm distribution

Division/District	Number of farms
Khulna	
Jessore	78
Khulna	92
Kushtia	79
Satkhira	80
Subtotal	329
Rajshahi	
Natore	87
Nawabgunj	85
Rajshahi	99
Subtotal	271
Total	600

However, in this book, for computational ease and brevity, the exact multistage block sampling procedure has been deviated. Here a subsample of 96 farms from seven selected districts of both divisions has been taken. The measurements of the variables used in this data are as follows:

(a) x_{ijk2}: Biological-Chemical (BC) Input:

This concludes both high yielding variety (HYV) and local variety of seeds, the chemical fertilizers, pesticides and insecticides. BC input per acre is measured in money terms, i.e. in thousands of Taka.

(b) x_{ijk3}: Human labour:

The data on human labour are given in adult man-days. This includes family as well as hired labour. In order to compute the wage-bill, family labour is assigned imputed value which is equal to the average of the wages of casual hired labour and permanently hired labour.

Table 6.3
X2: Artificially generated data for n=96, m=2, s=2, T=24

i	j	k	x_{ijk2}	x_{ijk3}	x_{ijk4}
1	1	1	102.44028	14.62891	14.32892
1	1	2	17.69850	38.22080	4.26574
1	1	3	45.39083	12.41084	10.72648
1	1	4	22.10381	15.73724	7.26575
1	1	5	86.54461	11.51338	18.44165
1	1	6	83.43267	15.56228	6.41979
1	1	7	6.06479	14.29658	8.27100
1	1	8	30.97777	8.77406	9.70438
1	1	9	29.61556	17.14609	12.08422
1	1	10	20.02457	18.50295	4.89055
1	1	11	19.88290	20.42256	5.26336
1	1	12	7.97451	4.22572	14.84995
1	1	13	193.92471	66.66033	11.02774
1	1	14	70.74133	44.00222	14.97263
1	1	15	24.13951	16.04665	4.72805
1	1	16	21.09276	10.30161	1.68796
1	1	17	23.76728	13.89932	5.78294
1	1	18	2.47052	27.68553	1.99841
1	1	19	20.71276	56.91961	8.11510
1	1	20	2.79513	22.05987	1.57264
1	1	21	102.08237	199.58296	0.47101
1	1	22	6.71692	22.43248	12.04968
1	1	23	71.28814	59.62470	19.62987
1	1	24	36.30734	40.63420	15.53035
1	2	1	2.43445	38.22654	5.97177
1	2	2	15.77600	4.90910	4.52484
1	2	3	19.44831	36.81002	5.93403
1	2	4	4.36060	22.74602	17.63734
1	2	5	65.02488	14.40334	13.28677
1	2	6	42.18986	27.28044	3.74893
1	2	7	16.43922	19.17658	14.65504
1	2	8	6.06940	34.57428	9.15907
1	2	9	49.92091	23.44600	8.05458
1	2	10	3.18748	8.62565	1.81427
1	2	11	3.55431	30.22713	10.14179
1	2	12	10.69643	37.75961	3.13892

Table 6.3 continued

i	j	k	x_{ijk2}	x_{ijk3}	x_{ijk4}
1	2	13	2.69417	6.92328	2.15610
1	2	14	13.33684	38.13948	19.83900
1	2	15	19.46718	82.74967	19.90378
1	2	16	6.42926	16.27630	14.12346
1	2	17	56.03015	22.88885	13.34708
1	2	18	8.53581	22.77424	9.83104
1	2	19	17.06023	17.07747	0.17146
1	2	20	22.57290	29.96200	2.71205
1	2	21	0.66291	30.19209	18.85556
1	2	22	40.69153	159.69408	14.34513
1	2	23	29.88689	3.33813	4.01511
1	2	24	79.88276	7.95492	19.61812
2	1	1	32.10366	25.13421	3.88157
2	1	2	65.93636	74.07070	16.12026
2	1	3	20.40541	19.44403	10.45981
2	1	4	4.70022	18.24024	8.92963
2	1	5	50.39289	21.80480	0.93681
2	1	6	16.15953	3.48069	4.88509
2	1	7	23.26128	5.70259	8.03407
2	1	8	30.43460	15.06313	5.25635
2	1	9	49.90344	2.17226	10.96155
2	1	10	9.07107	24.08743	3.51882
2	1	11	4.87983	12.93944	1.18945
2	1	12	51.13453	10.00375	14.45123
2	1	13	72.33007	37.02710	12.15640
2	1	14	12.46095	28.49646	10.75508
2	1	15	21.89569	33.75606	3.02889
2	1	16	10.86493	12.99299	2.59360
2	1	17	11.08796	26.97580	17.46648
2	1	18	13.62096	55.59037	12.61626
2	1	19	7.54104	15.10099	12.78103
2	1	20	55.03338	6.18224	12.92152
2	1	21	30.15136	65.37304	16.65537
2	1	22	11.66532	41.10583	9.76576
2	1	23	8.85764	30.37288	11.45818
2	1	24	15.40990	23.15244	5.40089
2	2	1	9.97408	6.71242	13.90643

161

Table 6.3 continued

i	j	k	x_{ijk2}	x_{ijk3}	x_{ijk4}
2	2	2	21.89000	7.84958	2.96579
2	2	3	54.53267	3.00991	4.13541
2	2	4	34.41491	4.58495	8.61873
2	2	5	32.83451	10.64361	6.88896
2	2	6	53.37914	7.32608	13.82809
2	2	7	30.39810	38.58833	12.28813
2	2	8	7.13028	11.13474	8.68824
2	2	9	55.80090	19.55713	8.22674
2	2	10	74.26427	20.56931	13.04514
2	2	11	62.25320	4.83799	14.62588
2	2	12	10.83466	19.09755	14.85305
2	2	13	16.75522	49.45385	5.30633
2	2	14	24.59812	89.69814	3.28932
2	2	15	6.45077	71.45801	9.85965
2	2	16	25.48525	25.43230	16.47815
2	2	17	2.57735	27.98616	9.36280
2	2	18	7.29960	6.18768	18.96755
2	2	19	36.23263	26.15473	9.84032
2	2	20	9.96431	95.91381	1.11511
2	2	21	17.41201	2.03546	19.78022
2	2	22	31.11064	15.47893	0.01891
2	2	23	329.25746	240.14670	6.50564
2	2	24	6.74175	39.45260	18.32898

Appendix 6B: Tabulation of results of the comparative study

Table 6.4
Powers of the LMMPI test for testing
H_0: $\rho_1 = \rho_2 = 0$, against H_a: $\rho_1 > 0$, $\rho_2 > 0$
(Figures in brackets are point optimal powers)

Case X1a, $N_i \times T = 2 \times 12$

	0.0	0.05	0.1	0.2	0.4
0.4	0.736	0.775	0.810	0.868	0.954
0.2	0.446	0.514	0.572	0.668	0.813
	(0.498)			(0.695)	
0.1	0.240	0.323	0.395	0.511	0.684
0.05	0.135	0.220	0.297	0.422	0.604
0.0	0.050	0.128	0.206	0.337	0.525
				(0.361)	
ρ_2/ρ_1	0.0	0.05	0.1	0.2	0.4

Case X1b, $N_i \times T = 3 \times 8$

	0.0	0.05	0.1	0.2	0.4
0.4	0.739	0.786	0.826	0.891	0.974
0.2	0.396	0.473	0.539	0.648	0.815
	(0.492)			(0.700)	
0.1	0.197	0.279	0.353	0.475	0.660
0.05	0.112	0.191	0.265	0.390	0.577
0.0	0.050	0.119	0.190	0.316	0.502
				(0.341)	
ρ_2/ρ_1	0.0	0.05	0.1	0.2	0.4

Case X1c, $N_i \times T = 2 \times 24$

	0.0	0.05	0.1	0.2	0.4
0.4	0.892	0.912	0.928	0.953	0.985
0.2	0.691	0.749	0.791	0.851	0.926
	(0.738)			(0.873)	
0.1	0.444	0.556	0.630	0.729	0.847
0.05	0.250	0.406	0.506	0.632	0.775
0.0	0.050	0.235	0.364	0.519	0.683
				(0.535)	
ρ_2/ρ_1	0.0	0.05	0.1	0.2	0.4

Table 6.4 continued

Case X1d, $N_i \times T = 3 \times 16$

0.4	0.926	0.943	0.957	0.976	0.996
0.2	0.691	0.754	0.800	0.865	0.945
	(0.758)			(0.894)	
0.1	0.402	0.513	0.593	0.703	0.841
0.05	0.209	0.345	0.444	0.579	0.742
0.0	0.050	0.186	0.297	0.450	0.628
				(0.478)	
ρ_2/ρ_1	0.0	0.05	0.1	0.2	0.4

Case X2a, $N_i \times T = 2 \times 12$

0.4	0.707	0.753	0.791	0.854	0.947
0.2	0.425	0.508	0.573	0.672	0.811
	(0.526)			(0.735)	
0.1	0.235	0.347	0.432	0.554	0.713
0.05	0.136	0.262	0.358	0.494	0.662
0.0	0.050	0.178	0.286	0.436	0.616
				(0.443)	
ρ_2/ρ_1	0.0	0.05	0.1	0.2	0.4

Case X2b, $N_i \times T = 2 \times 12$

0.4	0.646	0.712	0.764	0.847	0.958
0.2	0.332	0.439	0.518	0.635	0.794
	(0.511)			(0.747)	
0.1	0.176	0.302	0.397	0.529	0.696
0.05	0.107	0.239	0.340	0.481	0.654
0.0	0.050	0.179	0.286	0.437	0.616
				(0.443)	
ρ_2/ρ_1	0.0	0.05	0.1	0.2	0.4

Table 6.5

Calculated values of ρ_{10}^* and c_α^* for the POI-tests s(0.1,0.1), s(0.25,0.25) and c_α for the LM1 and LM2 tests at the five percent significance level for selected value of n, m, s and T

Design Matrix	$s(0.1, 0.1)$ test ρ_{10}^*	c_α^*	$s(o.25, 0.25)$ test ρ_{10}^*	c_α^*	LM1 test c_α	LM2 test c_α	n	m	s	T
X1(n x 3)	0.1279	0.9816	0.3559	1.0786	1.4947	3.5528	48	2	3	8
X1(n x 3)	0.1252	0.9591	0.3298	0.9924	1.2966	3.9919	24	2	3	4
X2(n x 3)	0.1222	0.9964	0.3441	1.1103	1.2743	3.7645	64	2	4	8
X2(n x 3)	0.1208	0.9691	0.3232	1.0112	1.0655	3.8244	24	2	3	4
X3(n x 3)	0.1313	0.9987	0.3693	1.1139	1.5421	3.6574	72	2	3	12
X3(n x 3)	0.1192	0.9579	0.3175	0.9950	1.4068	4.0323	24	2	3	4

Table 6.6

Estimated sizes of LM1 and LM2 tests for testing ρ_2 in the presence of ρ_1 using asymptotic critical values at 5% nominal level, for n=24 and p=3.

		Test Statistics	
Data Matrices	ρ_1	*LM1**	*LM2*
	0.0	.015	.059
X1	0.05	.018	.055
	0.1	.016	.050
	0.2	.016	.047
	0.3	.016	.038*
	0.4	.014	.031*
	0.5	.013	.026*
X2	0.0	.013	.048
	0.05	.012	.043
	0.1	.013	.040*
	0.2	.012	.032*
	0.3	.012	.030*
	0.4	.012	.028*
	0.5	.010	.024*
X3	0.0	.021	.059
	0.05	.023	.050
	0.1	.023	.045
	0.2	.022	.038*
	0.3	.019	.033*
	0.4	.019	.032*
	0.5	.018	.031*

* Indicates the sizes are significantly different than the nominal size of five percent.

Table 6.7
Estimated sizes of LM1 and LM2 tests for testing ρ_2 in the presence of ρ_1 using asymptotic critical values at 5% nominal level, for p=3

Data Matrices	Sample sizes	ρ_1	Test Statistics	
			LM1*	LM2*
X1	48	0.0	.019	.033
(n×3)		0.05	.020	.029
		0.1	.021	.031
		0.2	.021	.029
		0.3	.022	.028
		0.4	.023	.028
		0.5	.023	.027
X2	64	0.0	.019	.047**
(n×3)		0.05	.018	.038
		0.1	.018	.032
		0.2	.016	.028
		0.3	.016	.025
		0.4	.016	.023
		0.5	.017	.022
X3	72	0.0	.025	.042**
(n×3)		0.05	.026	.034
		0.1	.025	.030
		0.2	.028	.033
		0.3	.025	.030
		0.4	.025	.028
		0.5	.023	.024

** Indicates exceptional cases, when the size of the LM2 test differ insignificantly from the nominal size of five percent.

Table 6.8
Calculated values of the power function for X1 with n=24, m=2, s=3,
T=4 and p=3, using five percent critical values for testing

$$H_0: \rho_2 = 0,\ \rho_1 > 0,\ \text{against } H_a: \rho_1 > 0,\ \rho_2 > 0$$

Tests	$\rho_2 \backslash \rho_1 =$	0.0	0.05	0.1	0.2	0.3	0.4	0.5
s(0.1,0.1)	0.0	.050	.050	.050	.050	.050	.050	.050
s(0.25,0.25)		.050	.050	.050	.050	.050	.050	.050
LM1		.043	.048	.050	.049	.047	.043	.041*
LM2		.050	.045	.041	.038	.033	.026	.024
s(0.1,0.1)	0.05	.081	.082	.084	.089	.095	.103	.115
s(0.25,0.25)		.081	.082	.084	.089	.095	.104	.117
LM1		.074	.081	.080	.080	.082*	.086*	.095*
LM2		.049	.050	.047	.043	.044	.043	.038
s(0.1,0.1)	0.1	.119	.123	.128	.140	.155	.177	.208
s(0.25,0.25)		.118	.122	.127	.140	.156	.179	.213
LM1		.109	.112	.112*	.127*	.142*	.160*	.180*
LM2		.056	.056	.056	.056	.061	.067	.078
s(0.1,0.1)	0.2	.217	.228	.241	.271	.311	.365	.442
s(0.25,0.25)		.216	.227	.240	.272	.315	.374	.459
LM1		.198*	.201*	.217*	.247*	.277*	.311*	.370*
LM2		.100	.104	.110	.129	.147	.171	.217
s(0.1,0.1)	0.3	.335	.354	.375	.426	.490	.574	.683
s(0.25,0.25)		.335	.354	.377	.431	.501	.593	.714
LM1		.308*	.318*	.337*	.384*	.440*	.510*	.593*
LM2		.182	.190	.204	.241	.277	.325	.420
s(0.1,0.1)	0.4	.464	.490	.519	.586	.668	.766	.876
s(0.25,0.25)		.465	.493	.524	.596	.685	.792	.911
LM1		.434*	.458*	.484*	.537*	.607*	.684*	.788*
LM2		.270	.292	.315	.369	.432	.529	.646
s(0.1,0.1)	0.5	.594	.625	.659	.735	.821	.910	.976
s(0.25,0.25)		.598	.631	.668	.750	.843	.937	.998
LM1		.565*	.587*	.619*	.680*	.751*	.843*	.932*
LM2		.391	.416	.448	.523	.615	.722	.853

*Indicates the difference in power between POI and LM1 test is significant,
whereas the values without * shows that the differencein power between POI
and LM1 test is not significant.

Table 6.9
Calculated values of the power function for X2, with n=24, m=2, s=3, T=4 and p=3, using five percent critical values for testing

$$H_0: \rho_2 = 0, \rho_1 > 0, \text{ against } H_a: \rho_1 > 0, \rho_2 > 0$$

Tests	$\rho_2 \backslash \rho_1 =$	0.0	0.05	0.1	0.2	0.3	0.4	0.5
s(0.1,0.1)	0.0	.050	.050	.050	.050	.050	.050	.050
s(0.25,0.25)		.050	.050	.050	.050	.050	.050	.050
LM1		.044	.050	.049	.050	.050	.048	.049
LM2		.050	.043	.041	.032	.030	.029	.024
s(0.1,0.1)	0.05	.079	.081	.083	.087	.094	.102	.115
s(0.25,0.25)		.079	.080	.082	.087	.093	.102	.114
LM1		.074	.080	.084	.089	.090	.093	.102
LM2		.040	.041	.034	.033	.027	.029	.031
s(0.1,0.1)	0.1	.115	.119	.124	.136	.151	.173	.205
s(0.25,0.25)		.114	.119	.123	.135	.150	.173	.206
LM1		.107	.113	.123	.132	.148	.162	.185*
LM2		.050	.047	.046	.043	.043	.050	.060
s(0.1,0.1)	0.2	.206	.217	.229	.259	.299	.354	.435
s(0.25,0.25)		.208	.217	.229	.260	.301	.358	.442
LM1		.204	.214	.228	.256	.281*	.320*	.381*
LM2		.087	.088	.093	.105	.124	.145	.187
s(0.1,0.1)	0.3	.316	.334	.355	.405	.471	.560	.681
s(0.25,0.25)		.318	.337	.358	.411	.479	.572	.698
LM1		.304	.323	.339	.387*	.436*	.502*	.599*
LM2		.150	.156	.169	.198	.235	.292	.378
s(0.1,0.1)	0.4	.436	.463	.492	.561	.648	.759	.893
s(0.25,0.25)		.443	.470	.500	.572	.663	.776	.907
LM1		.430	.453	.479	.531*	.602*	.685*	.723*
LM2		.234	.254	.261	.321	.379	.468	.602
s(0.1,0.1)	0.5	.561	.594	.630	.713	.812	.922	1.000
s(0.25,0.25)		.572	.606	.643	.728	.827	.934	1.000
LM1		.555	.576*	.606*	.668*	.748*	.856*	.947*
LM2		.342	.368	.397	.463	.544	.675	.851

* Indicates the difference in power between POI and LM1 test is significant, whereas the values without * indicates that the difference in power between POI and LM1 test is not significant.

169

Table 6.10
Calculated values of the power function for X3, with n=24, m=2, s=3, T=4, and p=3 using five percent ritical values for testing
$$H_0: \rho_2 = 0, \rho_1 > 0, \text{ against } H_a: \rho_1 > 0, \rho_2 > 0$$

Tests	$\rho_2 \backslash \rho_1=$	0.0	0.05	0.1	0.2	0.3	0.4	0.5
s(0.1,0.1)	0.0	.050	.050	.050	.050	.050	.050	.050
s(0.25,0.25)		.050	.050	.050	.050	.050	.050	.050
LM1		.044	.049	.050	.048	.047	.045	.046
LM2		.050	.043	.039	.034	.029	.028	.025
s(0.1,0.1)	0.05	.081	.083	.085	.090	.097	.106	.119
s(0.25,0.25)		.081	.083	.085	.090	.097	.106	.120
LM1		.077	.795	.081	.083	.088	.096	.104*
LM2		.045	.042	.042	.041	.040	.039	.040
s(0.1,0.1)	0.1	.122	.126	.131	.144	.161	.184	.219
s(0.25,0.25)		.121	.125	.131	.144	.161	.186	.223
LM1		.115	.118	.121	.137	.154	.175	.199*
LM2		.067	.061	.061	.065	.072	.084	.099
s(0.1,0.1)	0.2	.225	.237	.250	.283	.327	.387	.472
s(0.25,0.25)		.224	.236	.250	.284	.330	.393	.484
LM1		.219	.239	.250	.274	.314	.357*	.419*
LM2		.120	.128	.139	.164	.184	.218	.273
s(0.1,0.1)	0.3	.350	.370	.393	.448	.518	.608	.725
s(0.25,0.25)		.349	.370	.394	.451	.525	.620	.744
LM1		.334	.354	.377	.421	.484*	.560*	.652*
LM2		.215	.235	.250	.289	.335	.403	.499
s(0.1,0.1)	0.4	.485	.513	.544	.615	.701	.803	.912
s(0.25,0.25)		.486	.515	.547	.621	.712	.818	.930
LM1		.476	.502	.531	.584*	.657*	.740*	.850*
LM2		.332	.357	.375	.436	.518	.617	.744
s(0.1,0.1)	0.5	.619	.652	.688	.766	.853	.939	.994
s(0.25,0.25)		.622	.656	.693	.775	.864	.951	1.000
LM1		.595	.623*	.656*	.735*	.807*	.899*	.969*
LM2		.464	.494	.528	.603	.701	.813	.931

* Indicates the difference in power between POI and LM1 test is significant, whereas the values without * indicates hat the difference in power between POI and LM1 test is not significant.

Table 6.11
Calculated values of the power function for X1, with n=48, m=2, s=3, T=8 and p=3, using five percent critical values for testing

$$H_0: \rho_2 = 0, \rho_1 > 0, \text{ against } H_a: \rho_1 > 0, \rho_2 > 0$$

Tests	$\rho_2 \backslash \rho_1=$	0.0	0.05	0.1	0.2	0.3	0.4	0.5
s(0.1,0.1)	0.0	.050	.050	.050	.050	.050	.050	.050
s(0.25,0.25)		.050	.050	.050	.050	.050	.050	.050
LM1		.042	.045	.048	.050	.050	.049	.045
LM2		.050	.039	.037	.039	.033	.034	.031
s(0.1,0.1)	0.05	.135	.140	.146	.161	.180	.206	.243
s(0.25,0.25)		.135	.140	.146	.161	.180	.207	.244
LM1		.126	.134	.139	.154	.171	.189*	.216*
LM2		.093	.103	.103	.111	.116	.131	.160
s(0.1,0.1)	0.1	.245	.257	.270	.302	.342	.395	.465
s(0.25,0.25)		.244	.256	.270	.302	.343	.396	.468
LM1		.245	.267	.280	.299	.330	.367*	.417*
LM2		.180	.198	.204	.228	.260	.290	.341
s(0.1,0.1)	0.2	.469	.490	.513	.565	.626	.697	.778
s(0.25,0.25)		.468	.490	.513	.566	.627	.699	.781
LM1		.484	.489	.521	.553	.601*	.654*	.715*
LM2		.405	.427	.439	.476	.518	.578	.650
s(0.1,0.1)	0.3	.651	.674	.699	.752	.809	.868	.926
s(0.25,0.25)		.650	.674	.699	.753	.811	.871	.929
LM1		.661	.689	.701	.733*	.784*	.834*	.886*
LM2		.593	.613	.632	.683	.728	.790	.846
s(0.1,0.1)	0.4	.781	.803	.825	.870	.913	.954	.985
s(0.25,0.25)		.782	.803	.826	.871	.915	.955	.987
LM1		.790	.800	.823	.856	.889*	.923*	.955*
LM2		.742	.759	.777	.819	.857	.900	.940
s(0.1,0.1)	0.5	.870	.888	.906	.939	.968	.990	1.000
s(0.25,0.25)		.871	.889	.906	.940	.969	.992	1.000
LM1		.873	.881	.892*	.921*	.945*	.968*	.993*
LM2		.843	.851	.870	.898	.928	.960	.990

* Indicates the difference in power between POI and LM1 test is significant, whereas the values without * indicates that the difference in power between POI and LM1 is not significant.

Table 6.12
Calculated values of the power function for X2, with n=64, m=2, s=4, T=8 and p=3, using five percent critical values for testing

$$H_0: \rho_2 = 0, \rho_1 > 0, \text{ against } H_a: \rho_1 > 0, \rho_2 > 0$$

Tests	$\rho_2\backslash\rho_1=$	0.0	0.05	0.1	0.2	0.3	0.4	0.5
s(0.1,0.1)	0.0	.050	.050	.050	.050	.050	.050	.050
s(0.25,0.25)		.050	.050	.050	.050	.050	.050	.050
LM1		.045	.050	.049	.047	.045	.047	.046
LM2		.050	.039	.035	.029	.028	.025	.024
s(0.1,0.1)	0.05	.150	.157	.164	.182	.205	.237	.283
s(0.25,0.25)		.149	.156	.163	.180	.204	.236	.282
LM1		.144	.154	.154	.168	.187	.211*	.250*
LM2		.083	.081	.080	.083	.088	.101	.127
s(0.1,0.1)	0.1	.284	.298	.315	.353	.402	.465	.548
s(0.25,0.25)		.282	.297	.313	.352	.401	.465	.548
LM1		.271	.284	.304	.345	.375*	.424*	.492*
LM2		.159	.171	.179	.202	.224	.273	.333
s(0.1,0.1)	0.2	.548	.572	.598	.655	.719	.790	.864
s(0.25,0.25)		.548	.572	.598	.656	.721	.792	.866
LM1		.553	.558	.579	.628*	.673*	.739*	.804*
LM2		.402	.416	.443	.448	.558	.618	.707
s(0.1,0.1)	0.3	.742	.766	.790	.839	.888	.934	.972
s(0.25,0.25)		.744	.768	.792	.842	.891	.936	.973
LM1		.739	.753	.767*	.809*	.854*	.906*	.947*
LM2		.619	.639	.658	.707	.762	.833	.900
s(0.1,0.1)	0.4	.864	.882	.900	.934	.963	.985	.997
s(0.25,0.25)		.866	.885	.903	.936	.965	.986	.998
LM1		.849	.859*	.878*	.913*	.946*	.971*	.988*
LM2		.771	.790	.815	.850	.900	.944	.977
s(0.1,0.1)	0.5	.935	.947	.958	.978	.991	.999	1.000
s(0.25,0.25)		.936	.948	.960	.979	.992	.999	1.000
LM1		.920*	.934*	.944*	.965*	.981*	.993*	1.000
LM2		.874	.881	.902	.936	.967	.985	.998

* Indicates the difference in power between POI and LM1 test is significant, whereas the values without * indicates that the difference in power between POI and LM1 test is not significant.

Table 6.13
Calculated values of the power function for X3, with n=72, m=2, s=3, T=12 and p=3, using five percent critical values for testing

H_0: $\rho_2 = 0$, $\rho_1 > 0$, against H_1: $\rho_1 > 0$, $\rho_2 > 0$.

Tests	$\rho_2 \backslash \rho_1 =$	0.0	0.05	0.1	0.2	0.3	0.4	0.5
s(0.1,0.1)	0.0	.050	.050	.050	.050	.050	.050	.050
s(0.25,0.25)		.050	.050	.050	.050	.050	.050	.050
LM1		.040	.050	.043	.048	.040	.041	.044
LM2		.050	.043	.037	.037	.033	.031	.028
s(0.1,0.1)	0.05	.190	.198	.208	.231	.261	.300	.354
s(0.25,0.25)		.190	.198	.208	.231	.261	.300	.354
LM1		.190	.201	.198	.214	.235	.266	.310*
LM2		.142	.152	.152	.161	.174	.192	.233
s(0.1,0.1)	0.1	.354	.371	.389	.431	.482	.544	.620
s(0.25,0.25)		.354	.371	.389	.431	.482	.544	.620
LM1		.357	.376	.381	.411	.448*	.493*	.548*
LM2		.293	.309	.312	.337	.373	.423	.480
s(0.1,0.1)	0.2	.620	.642	.664	.712	.765	.820	.878
s(0.25,0.25)		.620	.641	.664	.712	.765	.821	.878
LM1		.619	.644	.659	.693	.743*	.778*	.825*
LM2		.562	.575	.600	.631	.678	.733	.782
s(0.1,0.1)	0.3	.783	.802	.821	.859	.897	.934	.966
s(0.25,0.25)		.783	.802	.821	.859	.897	.934	.966
LM1		.780	.797	.808	.833*	.862*	.892*	.924*
LM2		.736	.755	.766	.799	.830	.863	.905
s(0.1,0.1)	0.4	.878	.893	.907	.934	.958	.979	.994
s(0.25,0.25)		.878	.893	.907	.934	.959	.979	.994
LM1		.870	.875	.886*	.904*	.927*	.951*	.977*
LM2		.837	.851	.862	.883	.908	.940	.970
s(0.1,0.1)	0.5	.934	.944	.954	.972	.986	.996	1.000
s(0.25,0.25)		.934	.944	.954	.972	.986	.996	1.000
LM1		.918	.921	.928*	.948*	.970*	.984*	.999
LM2		.899	.905	.914	.937	.960	.979	.998

* Indicates the difference in power between POI and LM1 test is significant, whereas the values without * indicates that the difference in power between POI and LM1 test is not significant.

7 Optimal testing for serial correlation in a large number of small samples

Introduction

It is common in business surveys, opinion polls and in other sorts of routine surveys that businessmen and sociologists always deal with quite a large number of small groups of observations, where the observations in each group are obtained under similar conditions and are therefore correlated. For example, in routine household surveys, duplicate and/or triplicate samples from the same locality might be taken, or in hospitals, during a large study, blood pressure measurements might be taken several times from the same patient. Under such situations, of course, there may exist autocorrelation within each group.

Recently, Cox and Solomon (1988) discussed the problem of testing for such an autocorrelation coefficient in large numbers of small samples. They were particularly concerned with group sample sizes of three and illustrated their suggested procedure by applying it to triple observations on pulse rates for 200 people. This chapter re-examined Cox and Solomon's problem of testing for a non-zero autocorrelation coefficient, ρ, when one has a number of very small samples.

The aim of this chapter then is to see if one can succeed to construct some optimal tests (discussed in previous chapters) for the problem in which there are m independent samples, each of size k. This chapter applies King and Hillier's (1985) LBI, King's (1987b) POI and King and Wu's (1990) LMMPI test procedures for testing ρ in this model. The plan of this chapter is as follows. In the following section, a general model for m independent samples of size k is given. LBI, POI and LMMPI tests are derived. The final section contains some concluding remarks.

The model and the tests

The model

Suppose there are m independent samples of k observations which are denoted by

$$y_{it}, \quad t = 1,\dots,k,; \quad i = 1,\dots,m;$$

in which each sample has unknown mean

$$E\left(y_{it}\right) = \mu_i, \quad i = 1,\dots,m.$$

Let $u_{it} = y_{it} - \mu_i$ denote deviations about these means. It is assumed that u_{it} follows the stationary AR(1) process

$$u_{it} = \rho u_{it-1} + \varepsilon_{it}$$

where $\varepsilon_{it} \sim IN\left(0,\sigma^2\right)$ so that

$$u(i) = \left(u_{i1},\dots,u_{ik}\right)' \sim N\left(0,\sigma^2\,\Sigma(\rho)\right)$$

where $\Sigma(\rho)$ is the $k \times k$ matrix

$$\Sigma(\rho) = \frac{1}{1-\rho^2}
\begin{bmatrix}
1 & \rho & \rho^2 \dots \rho^{k-1} \\
\rho & 1 & \rho & \vdots \\
\rho^2 & \rho & 1 & \vdots \\
 & & & \ddots \\
\vdots & & & 1 & \rho \\
\rho^{k-1} & \dots & & \rho & 1
\end{bmatrix}.$$

Thus $y(i) = \left(y_{i1},\dots,y_{ik}\right)' \sim N\left(\mu_i \ell, \sigma^2\,\Sigma(\rho)\right)$

where $\ell = (1,\dots,1)'$ or

$$y(i) = \mu_i \ell + u(i), \quad i = 1,\dots,m, \tag{7.1}$$

176

in which

$$u(i) \sim N\left(0, \sigma^2 \Sigma(\rho)\right).$$

If n=mk is the total number of observations and if we stack the y(i) and u(i) to form the nx1 vectors

y=(y(1)´, ..., y(m)´)´and u=(u(1)´, ..., u(m)´)´,

respectively, then

$$y = X\mu + u,$$ (7.2)

where X is the nxm matrix $X = I_m \otimes \ell$, $\otimes$ is the Kroneker product and $\mu = \left(\mu_1, ..., \mu_m\right)'$. Then

$$u \sim N\left(0, \sigma^2 \Delta(\rho)\right)$$ (7.3)

in which $\Lambda(\rho) - I_m \otimes \Sigma(\rho)$ is an nxn matrix. In what follows, model (7.1) and its matrix version (7.2) can be extended to include regressors if required without loss of generality.

Optimal testing

The main problem of interest in this section is to test $H_0: \rho = 0$ against $H_a: \rho > 0$ in model (7.2) under (7.3). Note that this testing problem is invariant under the group of transformations

$$y \rightarrow \gamma_0 y + X\gamma$$

where γ_0 is a positive scalar and γ is a px1 vector. Again the similar three optimal tests are being constructed, namely the LBI test, the POI test and the LMMPI tests. As one can note that these three tests have already been discussed in detail in the previous chapters.

Locally best invariant tests

Here the locally best invariant test for the above problem is to reject H_0 for small values of

177

$$r = \frac{e' A_0 e}{e' e}$$

where

$$A_0 = \left. \frac{-\partial \Delta(\rho)}{\partial \rho} \right|_{\rho=0}$$

and e is the vector of OLS residuals from model (7.2) which provides a LBI test against H_a.

Now, one can easily note that $\left. \dfrac{-\partial \Sigma(\rho)}{\partial \rho} \right|_{\rho=0} = A_1$ (say),

where

$$A_1 = \begin{bmatrix} 0 & -1 & 0 & \ldots & & 0 \\ -1 & 0 & -1 & \ldots & & 0 \\ 0 & -1 & 0 & & & \vdots \\ & & & \ddots & & \\ \vdots & & & & 0 & -1 \\ 0 & 0 & \ldots & & -1 & 0 \end{bmatrix}, \tag{7.4}$$

so that

$$A_0 = I_m \otimes A_1$$

and

$$\begin{aligned} r &= \left\{ \sum_{i=1}^{m} e(i)' A_1 e(i) \right\} \bigg/ e' e \\ &= -2 \sum_{i=1}^{m} \sum_{t=2}^{k} e_{it} e_{it-1} \bigg/ \sum_{i=1}^{m} \sum_{t=1}^{k} e_{it}^2 . \end{aligned} \tag{7.5}$$

The form of equation (7.5) is a ratio of quadratic forms in normal variables, so the critical values, c_α, may be determined by the usual standard numerical techniques used to calculate critical values of the DW statistic (as referred to in earlier chapters). Unfortunately, the distribution of (7.5) is a function of the design matrix X through M_X where $M_X = I_n - X(X'X)^{-1}X'$. However, bounds for c_α which are independent of X but depend on A_0 can be calculated in an analogous way to the familiar DW bounds (see King: 1987a, p. 28-29). Finally, methods of approximating the critical values of DW statistics can also be used to approximate c_α (see King: 1987a, p.25-27; and Evans and King:1985). Further, if one is interested in testing $H_0:\rho=0$, against the two-sided alternative $H_a':\rho \neq 0$, one can use the generalized Neyman-Pearson lemma (see Ferguson: 1967, p. 238) to obtain a locally best unbiased invariant (LBUI) test, given that the test's size and local unbiased conditions are satisfied.

Point optimal invariant tests

This subsection shows how to construct a point optimal invariant test of the null hypothesis $H_0:\rho=0$ against an alternative $H_a':\rho > 0$. To begin, one can consider the simpler problem of testing H_0 against the simple alternative hypothesis $H_1:\rho = \rho_1 > 0$, in (7.2) and (7.3) where ρ_1 is a known fixed point at which it is required to have optimal power in the alternative parameter space. Using King (1987b, equation 18), the POI test is to reject the null hypothesis for small values of

$$
\begin{aligned}
s(\rho_1) &= \frac{e'Ae}{e'e} \\
&= \frac{u'Au}{u'M_X u}
\end{aligned}
\tag{7.6}
$$

where $e = M_X y = M_X u$,

$$M_X = I_n - X(X'X)^{-1}X',$$

$$A = \Delta^{-1}(\rho_1) - \Delta^{-1}(\rho_1)X(X'\Delta^{-1}(\rho_1)X)^{-1}X'\Delta^{-1}(\rho_1).$$

The last equality of (7.6) is obtained by observing that $e'Ae = u'M_X A M_X u = u'Au$.

It is important to note that as $s(\rho_1)$ in (7.6), is a ratio of quadratic forms in normal variables and hence, like for the Durbin-Watson statistic, probabilities of the form

$$\Pr[s(\rho_1) < c_\alpha] = \Pr\left[\frac{u'Au}{u'M_xu} < c_\alpha\right]$$

$$= \Pr\left[u'(A - c_\alpha M_x)u < 0 | u \sim N(0, \Delta(\rho))\right]$$

$$= \Pr\left[\left(\Delta^{-1/2}(\rho_1)u\right)'\left(\Delta^{1/2}(\rho)\right)'\left(\Delta(\rho_1) - c_\alpha I_n\right)\Delta^{1/2}(\rho_1)\left(\Delta^{-1/2}(\rho_1)u\right) < 0\right]$$

$$(7.7)$$

can be found by computing

$$\Pr\left[\sum_{i=1}^{n}\lambda_i\xi_i^2 < 0\right], \qquad (7.8)$$

where $\lambda_1,...,\lambda_n$ are the eigenvalues of

$$\left(\Delta^{1/2}(\rho_1)\right)'\left(\Delta(\rho_1) - c_\alpha M_x\right)\Delta^{1/2}(\rho_1) \qquad (7.9)$$

and $\xi_1^2,...,\xi_n^2$ are independent chi-squared random variables each with one degree of freedom.

To find the exact critical values, (7.8) can be evaluated using either Koerts and Abrahamse's FQUAD subroutine, Farebrother's (1980) PAN procedure, Imhof's (1961) subroutine, or Davies' (1980) algorithm. A central question is how should ρ_1 be chosen? Strategies for choosing the point at which a POI test optimized power are discussed in detail in Chapter two where one approach suggested was to chose a ρ_1 value arbitrarily, such as $\rho_1=0.3$, -0.5 or 0.75 etc, choosen in Chapters three and five. Another is, to take the limit of $s(\rho_1)$ tests as ρ_1 tends to zero. This approach is called the LBI test approach as discussed above. There seems little point in optimizing power when it is very low (as the LBI test does) or when it is one or nearly one. Here Bhatti and King's (1990) approach is highly recommended, that is, optimizing power at a middle power value, say 0.5.

180

Alternative methods of computing (7.7) without first calculating eigenvalues have been briefly reviewed in Chapters five and six. As was noted in chapter five, Palm and Sneek's approach involves using householder transformations to tridiagonalize (7.9) whereas Shively et al's approach suggests the use of modified Kalman filter to calculate (7.7) in O(n) operations. For large sample sizes, this should result in considerable computational savings over the eigenvalue method.

A more complicated model and the LMMPI test

The error covariance matrix in (7.3) involves only one parameter, ρ, But in practice it may be that the structure of the autocorrelated observations change from sample to sample or from one group of observations to another group of observations. Examples of such groupings would include geographical location, industry, occupation and years of schooling, etc. This may cause the autocorrelation coefficient, ρ, to vary from sample to sample or from group to group. In such cases the covariance matrix $\Delta(\rho)$, in (7.3) can be expressed as,

$$\Delta(\bar{\rho}) = \begin{bmatrix} \Sigma(\rho_1) & 0 & \cdots\cdots & 0 \\ & & \ddots & \vdots \\ 0 & \Sigma(\rho_1) & \ddots & \vdots \\ \vdots & \ddots & \ddots & 0 \\ \vdots & & \ddots & \ddots \\ 0 & \cdots\cdots & 0 & \Sigma(\rho_m) \end{bmatrix}$$

$$= \bigoplus_{i=1}^{m} \Sigma(\rho_i), \tag{7.10}$$

which is in a block diagonal with submatrices of the form

$$\Sigma(\rho_i) = \left[(1-\rho_i)^2 I_k + \rho_i A_k + \rho_i(1-\rho_i)C_k \right]^{-1},$$

where I_k is the $k \times k$ identity matrix,

181

$$A_k = \begin{bmatrix} 1 & -1 & 0 & \cdots\cdots\cdots & 0 \\ -1 & 2 & -1 & \ddots & \vdots \\ 0 & -1 & 2 & \ddots & \ddots & \vdots \\ \vdots & \ddots & \ddots & \ddots & \vdots \\ \vdots & & \ddots & \ddots & -1 & 0 \\ \vdots & & & \ddots & 2 & -1 \\ 0 & \cdots\cdots\cdots & 0 & -1 & 1 \end{bmatrix}, \quad C_k = \text{diag}\,(1,0,...,1) \text{ and}$$

in (7.10), $\rho = (\rho_1, \rho_2, ..., \rho_m)'$.

The main objective of this subsection is to construct a one-sided test for testing whether these autocorrelation coefficients are equal to zero against the alternative that they have some positive values, i.e.,

$$H_0 : \rho_1 = \rho_2 = ..., \rho_m = 0$$

against

$$H_a : \rho_1 \geq 0, \ \rho_2 \geq 0, \ ..., \ \rho_m \geq 0,$$

with at least one inequality being a strict inequality. This is an m-dimensional hypothesis testing problem. Therefore King and Wu's (1990) theorem can be used in a similar way as in Chapter three to construct a LMMPI test for testing H_0 against H_a, which is to reject H_0 for small values of

$$d = \frac{e'Ae}{e'e}, \tag{7.11}$$

where e is the vector of OLS residuals from (7.2) as mentioned earlier and

$$A = \sum_{i=1}^{m} A_i$$

such that from (7.10)

$$A_i = - \left.\frac{\partial \Delta(\bar{\rho})}{\partial \rho_i}\right|_{\rho=0}, \quad i = 1,2,...,m,$$

182

$$
= \begin{bmatrix}
0 & 0 & \cdots & 0 & \cdots & \cdots & \cdots & \cdots & \cdots & \cdots & \cdots & \cdots & \cdots & \cdots & 0 \\
0 & 0 & & \vdots & & & & & & & & & & & \vdots \\
\vdots & & \ddots & \vdots & & & & & & & & & & & \vdots \\
0 & \cdots & \cdots & 0 & & & & & & & & & & & \vdots \\
\vdots & & & & \ddots & & & & & & & & & & \vdots \\
\vdots & & & & & 0 & -1 & 0 & \cdots & \cdots & 0 & & & & \vdots \\
\vdots & & & & & -1 & 0 & -1 & & & \vdots & & & & \vdots \\
\vdots & & & & & & & \ddots & & & \vdots & & & & \vdots \\
\vdots & & & & & 0 & -1 & 0 & & \ddots & \vdots & & & & \vdots \\
\vdots & & & & & \vdots & & & \ddots & & \vdots & & & & \vdots \\
\vdots & & & & & \vdots & & & & & 0 & -1 & & & \vdots \\
\vdots & & & & & 0 & \cdots & \cdots & \cdots & -1 & 0 & & & & \vdots \\
\vdots & & & & & & & & & & & \ddots & & & \vdots \\
\vdots & & & & & & & & & & & & 0 & 0 & \cdots & 0 \\
\vdots & & & & & & & & & & & & 0 & 0 & & \vdots \\
\vdots & & & & & & & & & & & & \vdots & & \ddots & \vdots \\
0 & \cdots & \cdots & \cdots & \cdots & \cdots & \cdots & \cdots & \cdots & & \cdots & 0 & \cdots & \cdots & 0
\end{bmatrix} \qquad (7.12)
$$

Thus by (7.10) and (7.12), a matrix A can be written as,

$$
A = I_m \otimes A_1,
$$

where A_1 is given by (7.4). Therefore, it is worth noting that (7.11) is also a LBI test of H_0 in the direction $\rho_1 = \rho_2 = ... = \rho_m > 0$. As it has been noticed in Chapters three, four and five.

Concluding remarks

This chapter has developed a precise optimal testing procedure for large numbers of small samples to detect serial correlation coefficient, ρ, of stationary first order autoregressive process in business and other applied sciences, particularly, statistics, econometrics and biometrics. In the previous section, three optimal tests, i.e. the LBI, POI and LMMPI tests have been derived.

In view of the calculations performed in the previous chapters, especially chapters five and six, there is good reason to believe that here also the POI test will perform reasonably well. In fact a number of Monte Carlo studies have already shown that POI test are typically more powerful than the LBI tests (see King: 1985a). From this work it can be surmised that POI tests will also be typically more powerful than LMMPI tests. The main interest in this chapter was not so much in the tedious exercise of presenting the proof that POI tests worked yet again but it provides an exercise of building and generalising the Cox and Solomon (1988) model and then it derives LBI, POI and LMMPI tests for their general model.

8 Summary and conclusion

In this book several aspects of hypothesis testing procedures associated with the linear regression model has been considered and then it has been applied to the multistage sample survey data. The principal aim of this book is to develop enhanced optimal tests (i.e. LB, LMMP, BO and PO) to ascertain their size and power properties and empirically compare the powers of the existing tests with those of the new optimal tests and the asymptotic LM tests. At the beginning of this study, optimal testing procedures were seen to be very tedious and cumbersome in choosing a power curve which would dominate the power of other tests. But the use of computer technology and the advances in numerical methods have helped with the choice of a test procedure which is most powerful for testing block effects and hence provides a more accurate decision in hypothesis testing. Of course the key to success was in understanding how to apply these procedures to the issues raised in chapter one.

In chapter two, the literature was reviewed for the readers to understand the issues and problems connected with sample surveys, regression analysis, optimal testing and existing tests for testing block effects. Chapter two also laid down the conditions under which the work would be done. In the same chapter a unified regression model was developed which could easily be used by both econometricians and statisticians. In the remaining chapter, the power properties of the optimal tests for testing block effects while dealing with the various models was explored.

The purpose of chapter three was to develop BO, PO and LMMP tests for testing equicorrelation coefficients in the two- and three-stage SSMN distribution models. Using the numerical iterative procedure suggested by King (1987b) and Davies and Harte (1987), the PE was computed and compared with the powers of the SenGupta's LB test and the two versions of the BO tests, namely, $r_{0.5}$ and $r_{0.8}$ at the five percent level of significance. It

185

was found that a BO test for the two-stage SSMN model is superior to SenGupta's LB test and also that it is approximately UMP. In chapter four, the analogous tests for the SMN distribution model were considered. It was found that the power and the critical values of the POI and LMMPI test for the case of the two-stage SMN model are easily obtainable from the standard F distribution. These tests are also UMPI. The POI and LMMPI tests for the case of the three-stage SMN model were derived and applied the POI test for the more complicated problem of testing subblock equicorrelation coefficient in the presence of main block equicorrelation.

The main purpose of chapter five was to consider a 2SLR model whose disturbances follow SMN distributions and to construct a POI test for testing block effects. The PE was computed and then compared with the powers of the POI and the existing tests from the King and Evans studies. It was found that the powers of the POI test is marginally better than those of the existing tests. It was also found that for some selected block sizes, the POI test is approximately UMPI and its approximate critical values, c_α^* are obtainable from the standard F distribution. It was observed that in 87 per cent of the cases the estimated sizes (based on c_α^*) was more than half of the nominal size. The LMMPI test was developed for testing multiparameters and it was found to be equivalent to Deaton and Irish's LM1 test.

In chapter six, the 3SLR model was considered and its contents mainly concentrated on the following two issues. Firstly, the LMMPI and POI tests were constructed for testing whether the block/subblock effects are equal to zero or they are positive. An evaluation of the power showed that the LMMPI test has quite good power for the data sets that was used in this experiment. This evaluation supports its usefulness. Secondly, a test for testing intra-subblock equicorrelation in the presence of intra-block equicorrelation developed was constructed. An exact POI test and asymptotic LM tests was constructed. The exact sizes and powers of the POI tests were computed using numerical iterative procedures, while the sizes and power of the LM tests were estimated using Monte Carlo simulation techniques. The estimated sizes of the LM1 test for the 100 per cent of the cases considered in this experiment which differ significantly than the nominal size of five percent. The sizes of the LM1 test have the tendancy to increase with the increase in the sample size whereas the sizes of the LM2 test decreases. Generally, the power pattern of the POI and LM1 tests are as similar as it was in the case of 2SLR model in Chapter five, i.e. the power of the PO1 test is marginally better than that of the LM1 test. The optimal testing procedures developed in chapter seven was concerned with testing for serial correlation in a large number of small samples by generalizing the work of Cox and Solomon (1986, 1988).

186

In conclusion it can be said that for testing block effects, if a problem satisfies the conditions for the use of an optimal testing procedure, one should be encouraged to use such a procedure. In a situation where no UMP test exists, a right choice from among optimal tests is important. In choosing a test one should always search for a test which through its power dominates the others in the alternative parameter space. This book further strengthens the claim of the superiority of the PO test in the area of testing for block effects because the powers of the PO tests are shown to be superior than those of the other existing tests. The main constraint on the generalised use of these tests are that they require a prior knowledge of the signs of the parameters under test. Obviously one can not and should not conclude despite this exercise that optimal tests are the perfect tests in all circumstances. There is still a need for a lot of research to be done but then, as was noticed, they are worthy of attention.

Bibliography

Amemiya, T. (1971), 'The Estimation of Variances in a Variance-Components Model', *International Economic Review*, 12, pp. 1-13.

Anderson, T.W. (1984), *An Introduction to Multivariate Statistical Analysis*, John Wiley, New York.

Ansley, C.F. (1979), 'An Algorithm for the Exact Likelihood of a Mixed Autoregressive-Moving Average Process', *Biometrika*, 66, pp. 59-65.

Ara, I. and King, M.L. (1993), 'Marginal Likelihood Based Tests of Regression Disturbances'. Paper presented at the 1993 meeting of the Australasian meeting of the Econometrics Society, University of Sydney, Sydney, Australia.

Balestra, P. and Nerlove, M. (1966), 'Pooling Cross Section and Time Series Data in the Estimation of a Dynamic Model The Demand for Natural Gas', *Econometrica*, 34, pp. 585-612.

Baltagi, B.H. (1981), 'Pooling: An Experimental Study of Alternative Testing and Estimation Procedures in a Two-Way Error Component Model', *Journal of Econometrics*, 17, pp. 21-49.

Baltagi, B.H., Chang, Y.J. and Li, Q. (1992), 'Monte Carlo Results on Several New and Existing Tests for the Error Component Model', *Journal of Econometrics*, 54, pp. 95-120.

Baltagi, B.H. and Li, Q. (1991), 'A Joint Test for Serial Correlation and Random Individual Effects,' *Statistics and Probability Letters*, vol 11, pp. 277-280.

Baltagi, B.H. and Raj, B. (1992), 'A Survey of Recent Theoretical Developments in the Econometrics of Panel Data,' *Empirical Economics*, vol 17, pp. 85-109.

Beach, C.M. and MacKinnon, J.G. (1978), 'A Maximum Likelihood Procedure for Regression with Autocorrelated Errors', *Econometrica*, 46, pp. 51-58.

Bhatti, M.I. (1992), 'Optimal Testing for Serial Correlation in Large Number of Small Samples', *Biometrical Journal*, Vol. 34, pp. 57-67.

Bhatti, M.I. (1993), 'Efficient Estimation of Random Ceofficient Models Based on Survey Data'. *Journal of Quantitative Economics*, Vol. 9, No. 1, pp. 99-110.

Bhatti, M.I. (1994), 'Optimal Testing for Equicorrelated Linear Regression Models', to appear in Statistical Papers.

Bhatti, M.I. (1995), 'A Ump-Invariant Test: An Example', *Journal of Applied Statistical Science*, Vol. 2, No.4.

Bhatti, M.I., and Barry, A.M. (1994), 'A Note on Testing Equicorrelation in Random Coefficient Models', *Technical Report No. STR. 1414-04*, Department of Statistics and Operation Research, KSU, Riyadh, Saudi Arabia.

Bhatti, M.I. and Czerkawski, C. (1993), 'The Determinant of Capital Structure in Japanese Corporate Sector', *International Journal of Business Studies*, Vol. 1(1), pp. 101-111.

Bhatti, M.I. and King, M.L. (1990), 'A Beta-Optimal Test of The Equicorrelation Coefficient', *Australian Journal of Statistics*, 32, pp. 87-97.

Bowley, A.L. (1906), 'Address to the Economic Science and Statistics Section of the British Association for the Advancement of Science', *Journal of the Royal Statistical Society*, 69, pp. 548-557.

Breusch, T.S. (1980), 'Useful Invariance Results for Generalized Regression Models', *Journal of Econometrics*, 13, pp. 327-340.

Breusch, T.S., and Pagan, A.R. (1980), 'The Lagrange Multiplier Test and its Applications to Model Specification in Econometrics', *Review of Economic Studies*, 47, pp. 239-253.

Brooks, R. (1993), 'Alternate Point Optimal Tests for Regression Coefficient Stability', *Journal of Econometrics*, 57, pp. 356-376.

Buse, A. (1980), 'Tests for Additive Heteroscedasticit: Some Preliminary Results', paper presented to the Fourth World Congress of the Econometric Society, Aix-en-Provencc.

Campbell, C. (1977), 'Properties of Ordinary and Weighted Least Squares Estimates of Regression Coefficients for Two-Stage Samples', in Proceedings of the Social Statistics Section, *American Statistical Association*, pp. 800-805.

Christensen, R. (1984), 'A Note on Ordinary Least Squares Methods for Two-Stage Sampling', *Journal of the American Statistical Association*, 79, pp. 720-721.

Christensen, R. (1986), 'Methods for the Analysis of Cluster Sampling Models', *Technical Report*, No. 7, pp. 22-86, Montana State University, Department of Mathematical Sciences.

Christensen, R. (1987a), 'The Analysis of Two-stage Sampling Data by Ordinary Least Squares', *Journal of the American Statistical Association*, 82, pp. 492-498.

Christensen, R. (1987b), *Plane Answers to Complex Questions: The Theory of Linear Models*, Springer-Verlag, New York.

Cochran, W.G. (1942), 'Sampling Theory When the Sampling Units are of Unequal Sizes', *Journal of the American Statistical Association*, 37, pp. 199-212.

Cochran, W.G. (1946), 'Relative Accuracy of Systematic and Stratified Random Samples for a Certain Class of Population', *Annals of Mathematical Statistics*, 17, pp. 164-177.

Cochran, W.G. (1953), *Sampling Techniques*, 1st edition, 2nd edition (1963), John Wiley, New York.

Cox, D.R. (1961), 'Tests of Separate Families of Hypotheses', Proceedings of the Fourth Berkeley Symposium on Mathematical Statistics and Probability, 1, University of California Press, Berkeley, pp. 105-123.

Cox, D.R. (1962), 'Further Results on Tests of Separate Families of Hypotheses', *Journal of the Royal Statistical Society* B, 24, pp. 405-424.

Cox, D.R. and Hinkley, D.V. (1974), *Theoretical Statistics*, Chapman and Hall, London.

Cox, D.R. and Solomon, P.J. (1986), 'Analysis of Variability with Large Numbers of Small Samples', *Biometrika*, 73, pp. 543-554.

Cox, D.R. and Solomon, P.J. (1988), 'On Testing for Serial Correlation in Large Numbers of Small Samples', *Biometrika*, 75, pp. 145-148.

Crowder, M.J. (1985), 'A Distributional Model for Repeated Failure Time Measurements', *Journal of the Royal Statistical Society*, B, 47, pp. 447-452.

Davies, R.B. (1969), 'Beta-Optimal Tests and an Application to the Summary Evaluation of Experiments', *Journal of the Royal Statistical Society*, B, 31, pp. 524-538.

Davies, R.B. (1980), 'Algorithm AS155. The Distribution of a Linear Combination of χ^2 Random Variables', *Applied Statistics*, 29, pp. 323-333.

Davies, R.B. and Harte, D.S. (1987), 'Tests for Hurst Effect', *Biometrika*, 74, pp. 95-101.

Deaton, A. and Irish, M. (1983), 'Block Effects in Regression Analysis Using Survey Data', unpublished manuscript.

Deming, W.E. (1950), *Some Theory of Sampling*, Wiley, New York.

Dickens, W.T. (1990), 'Error Components in Grouped Data: Is It Ever Worth Weighting?' *The Review of Economics and Statistics*, vol. 72, pp. 328-333.

Dorfman, A.H. (1993), 'A Comparison of Design-Based and Model-Based Estimators of the Finite Population Distribution Function, *Australian Journal of Statistics*, Vol. 35(1), pp. 29-41.

Efron, B. (1975), 'Defining the Curvature of a Statistical Problem (with Applications to Second Order Efficiency)', *Annals of Statistics*, 3, pp. 1189-1242.

Evans, M.A. and King, M.L. (1985), 'A Point Optimal Test for Heteroscedastic Disturbances', *Journal of Econometrics*, 27, pp. 163-178.

Farebrother, R.W. (1980), 'Algorithm AS153: Pan's Procedure for the Tail Probabilities of the Durbin-Watson Statistic', *Applied Statistics*, 29, pp. 224-227 and 30, pp. 189.

Ferguson, T.S. (1967), *Mathematical Statistics A Decision Theoretic Approach*, Academic Press, New York.

Fisher, R.A. (1925), 'Theory of Statistical Estimation', Proceedings of the Cambridge Philosophical Society, 122, pp. 700-725.

Fuller, W.A. (1975), 'Regression Analysis for Sample Surveys', Shankhya, C, 37, pp. 117-132.

Fuller, W.A. and Battese, G.E. (1973), 'Transformations for Estimation of Linear Models with Nested Error Structures', *Journal of the American Statistical Association*, 68, pp. 626-632.

Fuller, W.A. and Battese, G.E. (1974), 'Estimation of Linear Models with Cross-Error Structure', *Journal of Econometrics*, 2, pp. 67-78.

Fuller, W.A., Kennedy, W., Schnell, D., Sullivan, G. and Park, H.J. (1986), 'PC CARP Ames', Iowa State University Laboratory.

Geisser, S. (1963), 'Multivariate Analysis of Variance for a Special Covariance Case', *Journal of the American Statistical Association*, 58, pp. 660-669.

Godambe, V.P. (1966), 'A New Approach to Sampling from Finite Populations', *Journal of the Royal Statistical Society*, B, 28, pp. 310-328.

Godfrey, L.G. (1988), *Misspecification Tests in Econometrics: The Lagrange Multiplier Principle and other Approaches*, Cambridge University Press, Cambridge.

Goldfeld, S.M. and Quandt, R.E. (1965), 'Some Tests of Heteroscedasticity', *Journal of the American Statistical Association*, 60, pp. 539-547.

Greenwald, B.C. (1983), 'A General Analysis of the Bias in the Estimated Standard Errors of Least Squares Coefficients', *Journal of Econometrics*, 22, pp. 323-338.

Graybill, F.A. (1969), *Introduction to Matrices with Applications in Statistics*, Wadsworthy, Belmont, California.

Halperin, M. (1951), 'Normal Regression Theory in the Presence of Intra-Class Correlation', *The Annals of Mathematical Statistics*, 22, pp. 573-580.

Hansen, M.H. and Hurwitz, W.N. (1949), 'On the Determination of the Optimum Probabilities In Sampling', *Annals of Mathematical Statistics*, 20, pp. 426-432.

Hansen, M.H., Hurwitz, W.N. and Madow, W.G. (1953), *Sampling Survey Methods and Theory*, Volumes I and II, John Wiley, New York.

Hartley, H.O. and Rao, J.N.K. (1968), 'A New Estimation Theory for Sample Survey', *Biometrika*, 55, pp. 547-557.

Hartley, H.O. and Rao, J.N.K. (1969), 'A New Estimation Theory for Sample Survey II, ' *In New Developments in Survey Sampling*, N.L. Johnson and H. Smith (Eds.), pp. 147-169, Wiley Interscience, New York,.

Harvey, A.C. and Phillips, G.D.A. (1974), 'A Comparison of the Power of Some Tests for Heteroscedasticity in the General Linear Model', *Journal of Econometrics*, 2, pp. 307-316.

Harrison, M.J. and McCabe, B.P.M. (1979), 'A New Test for Heteroscedasticity Based on Ordinary Least Squares Residuals', *Journal of the American Statistical Association*, 74, pp. 494-499.

Harrison, M.J. (1980), 'The Small Sample Performance of the Szroeter Bounds Tests for Heteroscedasticity and a Simple Test for Use When Szroeter's Test is Inconclusive', *Oxford Bulletin of Economic and Statistics*, 42, pp. 235-250.

Hausman, J.A. (1978), 'Specification Tests in Econometrics', *Econometrica*, 46, pp. 1251-1271.

Hidiroglou, M.A., Fuller, W.A. and Hickman, R.D. (1980), 'SUPER CARP (6th Edition)', Ames Iowa State University, Statistical Laboratory.

Hillier, G.H. (1987), 'Classes of Similar Regions and Their Power Properties for Some Econometric Testing Problems, ' *Econometric Theory*, 3, pp. 1-44.

Holt, M.M. (1977), 'SURREGR: Standard Errors of Regression Coefficients From Sample Survey Data', Research Triangle Park Research Triangle Institute.

Holt, D. and Scott, A.J. (1981), 'Regression Analysis Using Survey Data', *The Statistician*, 30, pp. 169-178.

Holt, D., Smith, T.M.F. and Winter, P.D. (1980), 'Regression Analysis of Data from Complex Surveys', *Journal of the Royal Statistical Society*, A., vol. 143, pp. 474-487.

Honda, Y. (1985), 'Testing the Error Components with Non-normal Disturbances', *Review of Economic Studies*, 52, pp. 681-690.

Honda, Y. (1989), 'On the Optimality of Some Tests of the Error Covariance Matrix in the Linear Regression Model', *Journal of the Royal Statistical Society*, Series B, Vol. 51, pp. 71-79.

Honda, Y. (1991) 'A Standardized Test for the Error Components Model with the Two-way Layout', *Economic Letters*, 37, pp.125-128.

Hoque, A. (1988), 'Farm Size and Economic-Allocative Efficieny in Bangladesh Agriculture', *Applied Economics*, 20, pp. 1353-1368.

Hoque, A. (1991), 'An Application and Test for a Random Coefficient Model in Bangladesh Agriculture', *Journal of Applied Econometrics*, 6, pp. 77-90.

Hsiao, C. (1986), *Analysis of Panel Data*, Cambridge University Press.

Imhof, P.J. (1961), 'Computing the Distribution of Quadratic Forms in Normal Variables', *Biometrika*, 48, pp. 419-426.

IMSL Math/Library (1989), *User's Manual*, Softcover Edition 1.1, Houston ISML Inc., U.S.A.

Isaacson, S.L. (1951), 'On the Theory of Unbiased Tests of Simple Statistical Hypotheses Specifying the Values of Two or More Parameters', *Annals of Mathematical Statistics*, 22, pp. 217-234.

Kendall, M.G. and Buckland, W.R. (1960), *A Dictionary of Statistical Terms*, 4th Edition, Logman Scientific and Technical Publication, England.

Kiaer, A.N. (1895), 'Observation et Experiences Concernant Des Denombrements Representatifs', Discussion Appears in Liv. 1, XC111-XCV11. Bulletin International Statistical Institute, 9, Liv. 2, pp. 176-183.

King, M.L. (1980), 'Robust Tests for Spherical Symmetry and Their Application to Least Squares Regression', *Annals of Statistics*, 8, pp. 1265-1271.

King, M.L. (1981), 'The Alternative Durbin-Watson Test: An Assessment Durbin and Watson's Choice of Test Statistic', *Journal of Econometrics*, 17, pp. 51-66.

King, M.L. (1982), 'A Locally Optimal Bounds Test for Autoregressive Disturbances', a paper presented at the European meeting of the Econometric Society, Dublin.

King, M.L. (1983), 'Testing for Moving Average Regression Disturbances', *Australian Journal of Statistics*, 25, pp. 23-34.

King, M.L. (1985a), 'A Point Optimal Test for Autoregressive Disturbances', *Journal of Econometrics*, 27, pp. 21-37.

King, M.L. (1985b), 'A Point Optimal Test for Moving Average Regression Disturbances', *Econometrics Theory*, 1, pp. 211-222.

King, M.L. (1986), 'Efficient Estimation and Testing of Regressions with a Serially Correlated Error Component', *Journal of Quantitative Economics*, 2, pp. 231-247.

King, M.L. (1987a), 'Testing for Autocorrelation in Linear Regression Models: A Survey'. In M.L. King and D.E.A. Giles, eds., *Specification Analysis in the Linear Regression Model*. Rutledge and Kegan Paul, London, pp. 19-73.

King, M.L. (1987b), 'Towards a Theory of Point Optimal Testing', *Econometric Reviews*, 6, pp. 169-218.

King, M.L. (1989), 'Testing for Fourth-Order Autocorrelation in Regression Disturbances when First-Order Autocorrelation is Present', *Journal of Econometrics*, 41, pp. 284-301.

King, M.L. and Evans, M.A. (1986), 'Testing for Block Effects in Regression Models Based on Survey Data', *Journal of the American Statistical Association*, 81, pp. 677-679.

King, M.L. and Evans, M.A. (1988), 'Locally Optimal Properties of the Durbin-Watson Test,' *Econometric Theory*, 4, pp. 509 516.

King, M.L. and Hillier, G.H. (1985), 'Locally Best Invariant Tests of the Error Covariance Matrix of the Linear Regression Model', *Journal of the Royal Statistical Society*, B, 47, pp. 98-102, *Econometrics*, 23, pp. 35-48.

King, M.L. and Giles, D.E.A. (1984), 'Autocorrelation and Pre-Testing in the Linear Model Estimation, Testing and Prediction', *Journal of Econometrics*, 25, pp. 35-48.

King, M.L. and Skeels, C.L. (1984), 'Joint Testing for Serial Correlation and Heteroscedasticity in the Linear Regression Model', Paper presented at the Australasian Meeting of the Econometrics Society, Sydney.

King, M.L. and Wu, X.P. (1990), 'Locally Optimal One-Sided Tests for Multiparameter Hypothesis'. Paper presented at the Sixth World Congress of the Econometric Society, Barcelona, Spain.

Kish, L. (1965), *Survey Sampling*, John Wiley, New York.

Kish, L. and Frankel, M.R. (1974), 'Inference from Complex Samples', *Journal of the Royal Statistical Society*, B, 36, pp. 1-37.

Kloek, T,(1981), 'OLS Estimation in a Model Where a Microvariable is Explained by Aggregates and Contemporaneous Disturbances are Equicorrelated', *Econometrica*, 49, pp. 205-207.

Koerts, J. and Abrahamse, A.P.J. (1969), *On the Theory and Application of the General Linear Model*. Rotterdam University Press, Rotterdam .

Konijn, H.S. (1962), 'Regression Analysis in Sample Surveys', *Journal of the American Statistical Association*, 57, pp. 590-606.

Körösi, G., Mátyás, L. and Székely, I. (1992), *Practical Econometrics*. Avebury, Aldershot, U.K.

195

Kott, P.S. (1991), 'A Model-Based Look at Linear Regression with Survey Data', *The American Statistician*, 45, 2, pp. 107-112.

Krämer, W. and Sonnberger, H. (1986), *The Linear Regression Model Under Test*, Physica-Verlag, Heidelberg.

Lehmann, E.L. (1959), *Testing Statistical Hypothesis*, 1st edition (2nd edition, 1986), John Wiley, New York.

Lehmann, E.L. and Stein, C. (1948), 'Most Powerful Tests of Composite Hypothesis, I. Normal Distributions', *Annals of Mathematical Statistics*, 19, pp. 495-515.

Maddala, G.S. (1971), 'The Use of Variance Components Models in Pooling Cross Section and Time Series Data', *Econometrica*, 39, pp. 341-358.

Maddala, G.S. (1977), *Econometrics*. Kogakusta, McGraw-Hill.

Madow, W.G. and Madow, L.H. (1944), 'On the Theory of Systematic Sampling', *Annals of Mathematical Statistics*, 20, pp. 333-354.

Magnus, J.R. (1978), 'Maximum Likelihood Estimation of the GLS Model with Unknown Parameters in the Disturbance Covariance Matrix', *Journal of Econometrics*, 7, pp. 281-312.

Martin, R.S., Reinsch, C. and Wilkinson, J.H. (1968), 'Householder's Tridiagonalization of a Symmetric Matrix', *Numerische Mathematik* 11, pp. 181-195.

Mátyás, L. and Sevestre, P. (1992), *The Econometrics of Panel Data*, Kluwer Academic Publishers.

Moulton, B.R. (1986), 'Random Group Effects and the Precision of Regression Estimates', *Journal of Econometrics*, 32, pp. 385-397.

Moulton, B.R. (1990), 'An Illustration of a Pitfall in Estimating the Effects of Aggregate Variables on Micro Units', *The Review of Economics and Statistics*, 72, pp. 334-338.

Moulton, B.R. and Randolph, W.C. (1989), 'Alternative Tests of the Error Components Model', *Econometrica*, 57, pp. 685-693.

Nathan, G. and Holt, D. (1980), 'The Effect of Survey Design on Regression Analysis', *Journal of the Royal Statistical Society*, B., 42, pp. 377-386.

Nerlove, M. (1971), 'A Note on Error Components Models', *Econometrica*, 39, pp. 383-396.

Newey, W.K. (1985), 'Maximum Likelihood Specification Testing and Conditional Moment Tests', *Econometrica*, 53, pp. 1047-1070.

Neyman, J. (1934), 'On the Two Different Aspects of the Representative Method: The Method of Stratified Sampling and the Method of Purposive Selection', *Journal of the Royal Statistical Society*, 97, pp. 559-606.

Neyman, J. (1935), 'Sur la Vérification des Hypothéses Statistiques Composées', Bulletin de la Société Mathematique de France, 36, pp. 346-366.

Neyman, J. and Pearson, E.S. (1936), 'Contributions to the Theory of Testing Statistical Hypothesis. I. Unbiased Critical Regions of Type A and Type A_1', *Statistical Research Memoirs*, 1, pp. 1-37.

Neyman, J. and Pearson, E.S. (1938), 'Contributions to the Theory of Testing Statistical Hypotheses II', *Statistical Research Memoirs*, 2, pp. 25-37.

Neyman, J. and Scott, E.L. (1967), 'On the Use of C(α),Optimal Tests of Composite Hypotheses', *Bulletin of the International Statistical Institute*, 41, pp. 477-497.

Pagan, A.R. and Wickens, M.R. (1989), 'A Survey of Some Recent Econometric Methods', *Economic Journal*, 99, pp. 962-1025.

Palm, F.C. and Sneek, J.M. (1984), 'Significance Tests and Spurious Correlation in Regression Models with Autocorrelated Errors', *Statistische Hefte*, 25, pp. 87-105.

Pfefferman, D. (1985), 'Regression Models for Grouped Population in Cross-Section Surveys', *International Statistical Review*, 53, pp. 37-59.

Prakash, S. (1979), 'Contributions to Bayesian Analysis in Heteroscedastic Models', unpublished Doctoral Dissertation (University of New England, Armidale, NSW, Australia).

Rao, C.R. (1962), 'Efficient Estimates and Optimum Inference Procedures in Large Samples', *Journal of the Royal Statistical Society*, B, 24, pp. 46-72.

Rao, C.R. (1963), 'Criteria of Estimation in Large Samples', Sankh $\bar{y}$ a, 25, pp. 189-206.

Rao, C.R. (1973), *Linear Statistical Inference and Its Applications.*, John Wiley, New York.

Rao, C.R., and Kleffe, J. (1980), 'Estimation of Variance Components', in *Handbook of Statistics*, Krishnakumar, P.R. (ed.), 1, pp. 1-40, North Holland.

Rao, J.N.K., Sutradhar, B.C. and Yue, K. (1991), 'Generalised Least Squares F-Test in Regression Analysis with Two-Stage Cluster Samples', Technical Report, Laboratory for Research in Statistics and Probability, Carleton University, Canada.

Rao, J.N.K., Sutradhar, B.C. and Yue, D. (1993), 'Generalised Least Square F-Test in Regression Analysis with Two-Stage Cluster Samples', *Journal of the American Statistical Association*, vol. 88, no. 424 pp. 1388-1391.

Royall, R.M. (1968), 'An Old Approach to Finite Population Sampling Theory', *Journal of the American Statistical Association*, 63, pp 1269-1279.

Royall, R.M. (1970), 'On Finite Population Sampling Theory Under Certain Linear Regression Models', *Biometrika*, 57, pp. 377-387.

Royall, R.M. and Cumberland, W.G. (1981), 'An Empirical Study of the Ratio Estimator and Estimators of its Variance', *Journal of the American Statistical Association*, 76, pp. 66-87.

Sampson, A.R. (1976), 'Stepwise BAN Estimators for Exponential Families with Multivariate Normal Applications', *Journal of Multivariate Analysis*, 6, pp. 167-175.

Sampson, A.R. (1978), 'Simple BAN Estimators of Correlations for Certain Multivariate Normal Models with Known Variances', *Journal of the American Statistical Association,* 73, pp. 859-862.

Särndal, C.E. (1978), 'Design Based and Model Based Inference in Survey Sampling', *Scandinavian Journal of Statistics*, 5, pp. 27-52.

Scott, A.J. and Holt, D. (1982), 'The Effects of Two-Stage Sampling on Ordinary Least Squares Methods', *Journal of the American Statistical Association*, 77, pp. 848-854.

Scott, A.J. and Smith, T.M.F. (1969), 'Estimation in Multistage Surveys', *Journal of the American Statistical Association*, 64, pp. 830-840.

Searle, S.R. (1971), *Linear Models.*, John Wiley, New York.

Searle, S.R. and Henderson, H.V. (1979), 'Dispersion Matrices for Variance Components Models', *Journal of the American Statistical Association,* 74, pp. 465-470.

Selliah, J.B. (1964), 'Estimation and Testing Problems in a Wishart Distribution', Technical Report No. 10, Stanford University.

SenGupta, A. (1981), 'Tests for Standardized Generalized Variances of Multivariate Normal Populations of Possibly Different Dimensions', Technical Report No. 50, Department of Statistics, Stanford University.

SenGupta, A. (1983), 'Generalized Canonical Variables', in Johnson, N.L. and Kotz, S. (eds.), *Encyclopaedia of Statistical Sciences*, 3, pp. 326-330, Wiley, New York.

SenGupta, A. (1987), 'On Tests for Equicorrelation Coefficient of a Standard Symmetric Multivariate Normal Distribution', *Australian Journal of Statistics*, 29, pp. 49-59.

SenGupta, A. (1988), 'On Loss of Power Under Additional Information -An Example', *Scandanavian Journal of Statistics*, 15, pp. 25-31.

SenGupta, A., and Vermeire, L. (1986), 'Locally Optimal Tests for Multiparameter Hypotheses', *Journal of the American Statistical Association*, 81, pp. 819-825.

Shively, T.S. (1988), 'An Exact Test for Stochastic Coefficient in a Time Series Regression Model', *Journal of Time Series Analysis*, 9, pp. 81-88.

Shively, T.S., Ansley, C.F. and Kohn, R. (1990), 'Fast Evaluation of the Distribution of the Durbin-Watson and Other Invariant Test Statistics in Time Series Regression', *Journal of the American Statistical Association*, 85, pp. 676-685.

Silvapulle, P. (1991), 'Point-Optimal Tests for Non-nested Time Series Models and Illustrations', unpublished Doctoral Dissertation, Monash University, Clayton.

Silvapulle, P. and King, M.L. (1991), 'Testing Moving Average Against Autoregressive Disturbances in the Linear Regression Model', *Journal of Business and Economic Statistics*, 9, pp. 329-335.

Sirivastava, M.S. (1965), 'Some Tests for the Intra-Class Correlation Model', *Annals of Mathematical Statistics*, 36, pp. 1802-1806.

Sukhatme, P.V. (1954), *Sampling Theory of Surveys, with Applications*, Ames, State College Press, Iowa.

Swamy, P.A.V.B. (1970), 'Efficient Inference in Random Coefficient Regression Models', *Econometrica*, 38. pp. 311 323.

Swamy, P.A.V.B. and Arora, S.S. (1972), 'The Exact Finite Sample Properties of the Estimators of Coefficients in the Error Components Regression Models', *Econometrica*, 40, pp. 253-260.

Tauchen, G. (1985), 'Diagnostic Testing and Evaluation of Maximum Likelihood Models', *Journal of Econometrica*, 30, pp. 415-443.

Volaw, D.F. (1948), 'Testing Compound Symmetry in a Normal Multivariate Distribution', *Annals of Mathematical Statistics*, 19, pp. 447-473.

Wallace, T.D. and Hussain, A. (1969), 'The Use of Error Components Models in Combining Cross Section with Time Series Data', *Econometrica*, 37, pp. 55-72.

Walsh, J.E. (1947), 'Concerning the Effect of Intra-Class Correlation on Certain Significance Tests', *Annals of Mathematical Statistics*, 18, pp. 88-96.

White, H. (1982), 'Maximum Likelihood Estimation of Misspecified Models', *Econometrica*, 50, pp. 1-25.

Wilks, S.S. (1946), 'Sample Criteria for Testing Equality of Means, Equality of Variances, and Equality of Covariances in a Normal Multivariate Distribution', *Annals of Mathematical Statistics*, 17, pp. 257-281.

Williams, P. and Sams, D. (1981), 'Household Headship in Australia: Further Developments to the IMPACT Projects's Econometric Model of Household Headship', IMPACT Project Research Centre Working Paper BP-26, University of Melbourne.

Williams, E.J. and Yip, P. (1989), 'Conditional Inference About an Equicorrelation Coefficient', *Australian Journal of Statistics*, 31, pp.138-142.

Wu, P.X. (1991), 'One-sided and Partially One-sided Multiparameter Hypothesis Testing in Econometrics', unpublished Doctoral Dissertations, Monash University, Clayton.

Wu, P.X. and Bhatti, M.I. (1994), 'Testing for Block Effects and Misspecification in Regression Models based on Survey Data', *Journal of Statistical Computation and Simulation*, Vol.50, Nos. 1-2, pp.75-90.

Wu, C.F.J., Holt, D. and Holmes, D.J. (1988), 'The Effect of Two-Stage Sampling on the F Statistics', *Journal of the American Statistical Association*, 83, pp. 150-159.

Yates, F. (1949), *Sampling Methods for Census and Surveys*, 1st Edition, 2nd Edition (1953); 3rd Edition (1969), Griffin, London.